# WASTE PRODUCTS AND POLLUTION

# WASTE PRODUCTS AND POLLUTION

*By*

**Dr. D.R. Khanna**
*Reader*
*Department of Zoology*
*Gurukul Kangri University*
*Haridwar (Uttarakhand)*
*(India)*

**DISCOVERY PUBLISHING HOUSE PVT. LTD.**
**NEW DELHI-110 002**

*Published by:*
**Tilak Wasan**
**DISCOVERY PUBLISHING HOUSE PVT. LTD.**
4383/4A, Ansari Road, Darya Ganj
New Delhi-110 002 (India)
*Phone* : +91-11-23279245, 43596064-65
*Fax* : +91-11-23253475
*E-mail* : parul.wasan@gmail.com
discoverypublishinghouse@gmail.com
*web* : www.discoverypublishinggroup.com

*First Edition:* **2012**

**ISBN: 978-93-5056-129-4**

**Waste Products and Pollution**



***Printed at:***
***Shree Balaji Art Press***
***Delhi***

# Preface

Waste (also known as rubbish, trash, refuse, garbage, or junk) is unwanted or unusable materials. Litter is waste which has been disposed of improperly, particularly waste which has been carelessly disposed of in plain sight, as opposed to waste which has been dumped to avoid paying for waste disposal fees.

In living organisms, waste is the unwanted substances or toxins that are expelled from them. More commonly, waste refers to the materials that are disposed of in a system of waste management.

Waste is directly linked to human development, both technologically and socially. The compositions of different wastes have varied over time and location, with industrial development and innovation being directly linked to waste materials. Examples of this include plastics and nuclear technology. Some components of waste have economical value and can be recycled once correctly recovered.

Waste is sometimes a subjective concept, because items that some people discard may have value to others. It is widely recognized that waste materials are a valuable resource, whilst there is debate as to how this value is best realized.

Pollution is the introduction of contaminants into a natural environment that causes instability, disorder, harm or discomfort to the ecosystem i.e. physical systems or living organisms. Pollution can take the form of chemical substances or energy, such as noise, heat, or light. Pollutants, the elements of pollution, can be foreign substances or energies, or naturally occurring; when naturally occurring, they are considered contaminants when they exceed natural levels. Pollution is often classed as point source or non-point source pollution. The Blacksmith Institute issues annually a list of the world's worst polluted places. In the 2007 issues the ten top nominees are located in Azerbaijan, China, India, Peru, Russia, Ukraine, and Zambia.

Air pollution has always been with us. According to a 1983 article in the journal *Science*, "soot found on ceilings of prehistoric caves provides ample evidence of the high levels of pollution that was associated with inadequate ventilation of open fires." The forging of metals appears to be a key turning point in the creation of significant air pollution levels outside the home. Core samples of glaciers in Greenland indicate increases in pollution associated with Greek, Roman and Chinese metal production.

"The solution to pollution is dilution", is a dictum which summarizes a traditional approach to pollution management whereby sufficiently diluted pollution is not harmful. It is well-suited to some other modern, locally scoped applications such as laboratory safety procedure and hazardous material release emergency management. But it assumes that the dilutant is in virtually unlimited supply for the application or that resulting dilutions are acceptable in all cases.

**Author**

# Contents

# Chapter 1

# Introduction

Waste (also known as rubbish, trash, refuse, garbage, or junk) is unwanted or unusable materials. Litter is waste which has been disposed of improperly, particularly waste which has been carelessly disposed of in plain sight, as opposed to waste which has been dumped to avoid paying for waste disposal fees.

In living organisms, waste is the unwanted substances or toxins that are expelled from them. More commonly, waste refers to the materials that are disposed of in a system of waste management.

Waste is directly linked to human development, both technologically and socially. The compositions of different wastes have varied over time and location, with industrial development and innovation being directly linked to waste materials. Examples of this include plastics and nuclear technology. Some components of waste have economical value and can be recycled once correctly recovered.

Waste is sometimes a subjective concept, because items that some people discard may have value to others. It is widely recognized that waste materials are a valuable resource, whilst there is debate as to how this value is best realized.

There are many waste types defined by modern systems of waste management, notably including:

- municipal solid waste (MSW)
- construction waste and demolition waste (C&D)

- institutional waste, commercial waste, and industrial waste (IC & I)
- medical waste (also known as clinical waste)
- hazardous waste, radioactive waste, and electronic waste
- biodegradable waste

### Proposed Definitions by Pongrácz and Pohjola (2004)

Non-wanted things created, not intended, or not avoided, with no purpose.

Things that were given a finite purpose thus destined to become useless after fulfilling it.

Things with well-defined purpose, but their performance ceased being acceptable.

Things with well-defined purpose, and acceptable performance, but their users failed to use them for the intended purpose.

Taiichi Ohno from Toyota Production System describes waste as "Any human activity that absorbs resources but creates no value".

### Reporting

There are many issues that surround reporting waste. It is most commonly measured by size or weight, and there is a stark difference between the two. For example, organic waste is much heavier when it is wet, and plastic or glass bottles can have different weights but be the same size. On a global scale it is difficult to report waste because countries have different definitions of waste and what falls into waste categories, as well as different ways of reporting. Based on incomplete reports from its parties, the Basel Convention estimated 338 million tonnes of waste was generated in 2001. For the same year, OCED estimated 4 billion tonnes from its member countries. Despite these inconsistencies, waste reporting is still useful on a small and large scale to determine key causes and locations, and to find ways of preventing, minimizing, recovering, treating, and disposing waste.

## COSTS

### Environmental Costs

Waste can attract rodents and insects which cause gastrointestinal parasites, yellow fever, worms, the plague and other conditions for humans. Exposure to hazardous wastes, particularly when they are burned, can cause various other diseases including cancers. Waste can contaminate surface water, groundwater, soil, and air which causes more problems for humans, other species, and ecosystems. Waste treatment and disposal produces significant green house gas (GHG) emissions, notably methane, which are contributing significantly to global climate change.

### Social Costs

Waste management is a significant environmental justice issue. Many of the environmental burdens cited above are more often borne by marginalized groups, such as racial minorities, women, and residents of developing nations. NIMBY (not-in-my-back-yard) is a popular term used to describes the opposition of residents to a proposal for a new development close to them. However, the need for expansion and siting of waste treatment and disposal facilities is increasing worldwide. There is now a growing market in the transboundary movement of waste, and although most waste that flows between countries goes between developed nations, a significant amount of waste is moved from developed to developing nations.

### Economic Costs

The economic costs of managing waste are high, and are often paid for by municipal governments. Money can often be saved with more efficiently designed collection routes, modifying vehicles, and with public education. Environmental policies such as pay as you throw can reduce the cost of management and reduce waste quantities. Waste recovery (that is, recycling, reuse) can curve economic costs because it avoids extracting raw materials and often cuts transportation costs. The location of waste treatment and disposal facilities often has an impact on property values due to noise, dust, pollution, unsightliness, and negative stigma. The informal waste sector consists mostly of waste pickers who scavenge for metals, glass, plastic, textiles, and other materials and then trade them for a profit. This sector can significantly alter or reduce waste in a particular system, but other negative economic effects come with the disease, poverty, exploitation, and abuse of its workers.

### Education and Awareness

Education and awareness in the area of waste and waste management is increasingly important from a global perspective of resource management. The Talloires Declaration is a declaration for sustainability concerned about the unprecedented scale and speed of environmental pollution and degradation, and the depletion of natural resources. Local, regional, and global air pollution; accumulation and distribution of toxic wastes; destruction and depletion of forests, soil, and water; depletion of the ozone layer and emission of 'green house' gases threaten the survival of humans and thousands of other living species, the integrity of the earth and its biodiversity, the security of nations, and the heritage of future generations. Several universities have implemented the Talloires Declaration by establishing environmental management and waste management programmes, e.g. the waste management universityproject. University and vocational education are promoted by various organizations, e.g. WAMITAB and Chartered Institution of Wastes Management.

## Pollution

Pollution is the introduction of contaminants into a natural environment that causes instability, disorder, harm or discomfort to the ecosystem i.e. physical systems or living organisms. Pollution can take the form of chemical substances or energy, such as noise, heat, or light. Pollutants, the elements of pollution, can be foreign substances or energies, or naturally occurring; when naturally occurring, they are considered contaminants when they exceed natural levels. Pollution is often classed as point source or non-point source pollution. The Blacksmith Institute issues annually a list of the world's worst polluted places. In the 2007 issues the ten top nominees are located in Azerbaijan, China, India, Peru, Russia, Ukraine, and Zambia.

Air pollution has always been with us. According to a 1983 article in the journal *Science,* "soot found on ceilings of prehistoric caves provides ample evidence of the high levels of pollution that was associated with inadequate ventilation of open fires." The forging of metals appears to be a key turning point in the creation of significant air pollution levels outside the home. Core samples of glaciers in Greenland indicate increases in pollution associated with Greek, Roman and Chinese metal production.

## Official Acknowledgement

The earliest known writings concerned with pollution were written between the 9th and 13th centuries by Persian scientists such as Muhammad ibn Zakariya Razi (Rhazes), Ibn Sina (Avicenna), and al-Masihi or were Arabic medical treatises written by physicians such as al-Kindi (Alkindus), Qusta ibn Luqa (Costa ben Luca), Ibn Al-Jazzar, al-Tamimi, Ali ibn Ridwan, Ibn Jumay, Isaac Israeli ben Solomon, Abd-el-latif, Ibn al-Quff, and Ibn al-Nafis. Their works covered a number of subjects related to pollution such as air contamination, water contamination, soil contamination, solid waste mishandling, and environmental assessments of certain localities.

King Edward I of England banned the burning of sea-coal by proclamation in London in 1272, after its smoke had become a problem. But the fuel was so common in England that this earliest of names for it was acquired because it could be carted away from some shores by the wheelbarrow. Air pollution would continue to be a problem in England, especially later during the industrial revolution, and extending into the recent past with the Great Smog of 1952. This same city also recorded one of the earlier extreme cases of water quality problems with the Great Stink on the Thames of 1858, which led to construction of the London sewerage system soon afterward.

It was the industrial revolution that gave birth to environmental pollution as we know it today. The emergence of great factories and consumption of immense quantities of coal and other fossil fuels gave rise to unprecedented air pollution and the large volume of industrial chemical discharges added to the growing load of untreated human waste. Chicago and Cincinnati were the first two American cities to enact laws ensuring cleaner air in 1881. Other cities followed around the country until early in the 20th century, when the short lived Office of Air Pollution was created under the Department of the Interior. Extreme smog events were experienced by the cities of Los Angeles and Donora, Pennsylvania in the late 1940s, serving as another public reminder.

Pollution became a popular issue after World War II, due to radioactive fallout from atomic warfare and testing. Then a non-nuclear event, The Great Smog of 1952 in London, killed at least 4000 people. This prompted some of the first major modern environmental legislation, The Clean Air Act of 1956.

Pollution began to draw major public attention in the United States between the mid-1950s and early 1970s, when Congress passed the Noise Control Act, the Clean Air Act, the Clean Water Act and the National Environmental Policy Act.

Bad bouts of local pollution helped increase consciousness. PCB dumping in the Hudson River resulted in a ban by the EPA on consumption of its fish in 1974. Long-term dioxin contamination at Love Canal starting in 1947 became a national news story in 1978 and led to the Superfund legislation of 1980. Legal proceedings in the 1990s helped bring to light Chromium-6 releases in California—the champions of whose victims became famous. The pollution of industrial land gave rise to the name brownfield, a term now common in city planning. DDT was banned in most of the developed world after the publication of Rachel Carson's *Silent Spring*.

The development of nuclear science introduced radioactive contamination, which can remain lethally radioactive for hundreds of thousands of years. Lake Karachay, named by the Worldwatch Institute as the 'most polluted spot' on earth, served as a disposal site for the Soviet Union thoroughout the 1950s and 1960s. Second place may go to the area of Chelyabinsk U.S.S.R. as the "Most polluted place on the planet".

Nuclear weapons continued to be tested in the Cold War, sometimes near inhabited areas, especially in the earlier stages of their development. The toll on the worst-affected populations and the growth since then in understanding about the critical threat to human health posed by radioactivity has also been a prohibitive complication associated with nuclear power. Though extreme care is practiced in that industry, the potential for disaster suggested by incidents such as those at Three Mile Island and Chernobyl pose a lingering specter of public mistrust. One legacy of nuclear testing before most forms were banned has been significantly raised levels of background radiation.

International catastrophes such as the wreck of the Amoco Cadiz oil tanker off the coast of Brittany in 1978 and the Bhopal disaster in 1984 have demonstrated the universality of such events and the scale on which efforts to address them needed to engage. The borderless nature of atmosphere and oceans inevitably resulted in the implication of pollution on a planetary level with the issue of global warming. Most recently the term persistent organic pollutant (POP) has come to describe a group of chemicals such as PBDEs and PFCs among others. Though their effects remain somewhat less well understood owing to a lack of experimental data, they have been detected in various ecological habitats far removed from industrial activity such as the Arctic, demonstrating diffusion and bioaccumulation after only a relatively brief period of widespread use.

Growing evidence of local and global pollution and an increasingly informed public over time have given rise to environmentalism and the environmental movement, which generally seek to limit human impact on the environment.

The major forms of pollution are listed below along with the particular pollutants relevant to each of them:

- Air pollution, the release of chemicals and particulates into the atmosphere. Common gaseous air pollutants include carbon monoxide, sulfur dioxide, chlorofluorocarbons (CFCs) and nitrogen oxides produced by industry and motor vehicles. Photochemical ozone and smog are created as nitrogen oxides and hydrocarbons react to sunlight. Particulate matter, or fine dust is characterized by their micrometre size $PM_{10}$ to $PM_{2.5}$.
- Light pollution, includes light trespass, over-illumination and astronomical interference.

**Littering**

- Noise pollution, which encompasses roadway noise, aircraft noise, industrial noise as well as high-intensity sonar.

- Soil contamination occurs when chemicals are released by spill or underground leakage. Among the most significant soil contaminants are hydrocarbons, heavy metals, MTBE, herbicides, pesticides and chlorinated hydrocarbons.
- Radioactive contamination, resulting from 20th century activities in atomic physics, such as nuclear power generation and nuclear weapons research, manufacture and deployment.
- Thermal pollution, is a temperature change in natural water bodies caused by human influence, such as use of water as coolant in a power plant.
- Visual pollution, which can refer to the presence of overhead power lines, motorway billboards, scarred landforms (as from strip mining), open storage of trash or municipal solid waste.
- Water pollution, by the release of waste products and contaminants into surface runoff into river drainage systems, leaching into groundwater, liquid spills, wastewater discharges, eutrophication and littering.

## Pollutants

A pollutant is a waste material that pollutes air, water or soil. Three factors determine the severity of a pollutant: its chemical nature, the concentration and the persistence.

## Sources and Causes

Air pollution comes from both natural and man made sources. Though globally man made pollutants from combustion, construction, mining, agriculture and warfare are increasingly significant in the air pollution equation.

Motor vehicle emissions are one of the leading causes of air pollution. China, United States, Russia, Mexico, and Japan are the world leaders in air pollution emissions. Principal stationary pollution sources include chemical plants, coal-fired power plants, oil refineries, petrochemical plants, nuclear waste disposal activity, incinerators, large livestock farms (dairy cows, pigs, poultry, etc.), PVC factories, metals production factories, plastics factories, and other heavy industry. Agricultural air pollution comes from contemporary practices which include clear felling and burning of natural vegetation as well as spraying of pesticides and herbicides.

About 400 million metric tons of hazardous wastes are generated each year. The United States alone produces about 250 million metric tons. Americans constitute less than 5 per cent of the world's population, but produce roughly 25 per cent of the world's $CO_2$, and generate approximately 30 per cent of world's waste. In 2007, China has overtaken the United States as the world's biggest producer of $CO_2$.

In February 2007, a report by the Intergovernmental Panel on Climate Change (IPCC), representing the work of 2,500 scientists from more than 130 countries, said that humans have been the primary cause of global warming since 1950. Humans have ways to cut greenhouse gas emissions and avoid the consequences of global warming, a major climate report concluded. But in order to change the climate, the transition from fossil fuels like coal and oil needs to occur within decades, according to the final report this year from the UN's Intergovernmental Panel on Climate Change (IPCC).

Some of the more common soil contaminants are chlorinated hydrocarbons (CFH), heavy metals (such as chromium, cadmium—found in rechargeable batteries, and lead—found in lead paint, aviation fuel and still in some countries, gasoline), MTBE, zinc, arsenic and benzene. In 2001 a series of press reports culminating in a book called *Fateful Harvest* unveiled a widespread practice of recycling industrial byproducts into fertilizer, resulting in the contamination of the soil with various metals. Ordinary municipal landfills are the source of many chemical substances entering the soil environment (and often groundwater), emanating from the wide variety of refuse accepted, especially substances illegally discarded there, or from pre-1970 landfills that may have been subject to little control in the U.S. or EU. There have also been some unusual releases of polychlorinated dibenzodioxins, commonly called *dioxins* for simplicity, such as TCDD.

Pollution can also be the consequence of a natural disaster. For example, hurricanes often involve water contamination from sewage, and petrochemical spills from ruptured boats or automobiles. Larger scale and environmental damage is not uncommon when coastal oil rigs or refineries are involved. Some sources of pollution, such as nuclear power plants or oil tankers, can produce widespread and potentially hazardous releases when accidents occur.

In the case of noise pollution the dominant source class is the motor vehicle, producing about ninety per cent of all unwanted noise worldwide.

Adverse air quality can kill many organisms including humans. Ozone pollution can cause respiratory disease, cardiovascular disease, throat inflammation, chest pain, and congestion. Water pollution causes approximately 14,000 deaths per day, mostly due to contamination of drinking water by untreated sewage in developing countries. An estimated 700 million Indians have no access to a proper toilet, and 1,000 Indian children die of diarrhoeal sickness every day. Nearly 500 million Chinese lack access to safe drinking water. 656,000 people die prematurely each year in China because of air pollution. In India, air pollution is believed to cause 527,700 fatalities a year. Studies have estimated that the number of people killed annually in the US could be over 50,000.

Oil spills can cause skin irritations and rashes. Noise pollution induces hearing loss, high blood pressure, stress, and sleep disturbance. Mercury has been linked to developmental deficits in children and neurologic symptoms. Older people are majorly exposed to diseases induced by air pollution. Those with heart or lung disorders are under additional risk. Children and infants are also at serious risk. Lead and other heavy metals have been shown to cause neurological problems. Chemical and radioactive substances can cause cancer and as well as birth defects.

**Environment**

Pollution has been found to be present widely in the environment. There are a number of effects of this:

- Biomagnification describes situations where toxins (such as heavy metals) may pass through trophic levels, becoming exponentially more concentrated in the process.
- Carbon dioxide emissions cause ocean acidification, the ongoing decrease in the pH of the Earth's oceans as $CO_2$ becomes dissolved.
- The emission of greenhouse gases leads to global warming which affects ecosystems in many ways.
- Invasive species can out compete native species and reduce biodiversity. Invasive plants can contribute debris and biomolecules ( allelopathy) that can alter soil and chemical compositions of an environment, often reducing native species competitiveness.
- Nitrogen oxides are removed from the air by rain and fertilise land which can change the species composition of ecosystems.
- Smog and haze can reduce the amount of sunlight received by plants to carry out photosynthesis and leads to the production of tropospheric ozone which damages plants.
- Soil can become infertile and unsuitable for plants. This will affect other organisms in the food web.
- Sulphur dioxide and nitrogen oxides can cause acid rain which lowers the pH value of soil.

**Environmental Health Information**

The Toxicology and Environmental Health Information Programme (TEHIP) at the United States National Library of Medicine (NLM) maintains a comprehensive toxicology and environmental health web site that includes access to resources produced by TEHIP and by other government agencies and organizations. This web site includes links to databases, bibliographies, tutorials, and other scientific and consumer-oriented resources. TEHIP also is responsible for the Toxicology Data Network (TOXNET) an integrated system of toxicology and environmental health databases that are available free of charge on the web.

TOXMAP is a Geographic Information System (GIS) that is part of TOXNET. TOXMAP uses maps of the United States to help users visually explore data from the United States Environmental Protection Agency's (EPA) Toxics Release Inventory and Superfund Basic Research Programmes.

Pollution control is a term used in environmental management. It means the control of emissions and effluents into air, water or soil. Without pollution control, the waste products from consumption, heating, agriculture, mining, manufacturing, transportation and other human activities, whether they accumulate or disperse, will degrade the environment. In the hierarchy of controls, pollution prevention and waste minimization are more desirable than pollution control.

The earliest precursor of pollution generated by life forms would have been a natural function of their existence. The attendant consequences on viability and population levels fell within the sphere of natural selection. These would have included the demise of a population locally or ultimately, species extinction. Processes that were untenable would have resulted in a new balance brought about by changes and adaptations. At the extremes, for any form of life, consideration of pollution is superseded by that of survival.

For humankind, the factor of technology is a distinguishing and critical consideration, both as an enabler and an additional source of byproducts. Short of survival, human concerns include the range from quality of life to health hazards. Since science holds experimental demonstration to be definitive, modern treatment of toxicity or environmental harm involves defining a level at which an effect is observable. Common examples of fields where practical measurement is crucial include automobile emissions control, industrial exposure (e.g. Occupational Safety and Health Administration (OSHA) PELs), toxicology (e.g. $LD_{50}$), and medicine (e.g. medication and radiation doses).

'The solution to pollution is dilution', is a dictum which summarizes a traditional approach to pollution management whereby sufficiently diluted pollution is not harmful. It is well-suited to some other modern, locally scoped applications such as laboratory safety procedure and hazardous material release emergency management. But it assumes that the dilutant is in virtually unlimited supply for the application or that resulting dilutions are acceptable in all cases.

Such simple treatment for environmental pollution on a wider scale might have had greater merit in earlier centuries when physical survival was often the highest imperative, human population and densities were lower, technologies were simpler and their byproducts more benign. But these are often no longer the case. Furthermore, advances have enabled measurement of concentrations not possible before. The use of statistical methods in evaluating outcomes has given currency to the principle of probable harm in cases where assessment is warranted but resorting to deterministic models is impractical or unfeasible. In addition, consideration of the environment beyond direct impact on human beings has gained prominence.

Yet in the absence of a superseding principle, this older approach predominates practices throughout the world. It is the basis by which to gauge concentrations of effluent for legal release, exceeding which penalties are assessed or restrictions applied. The regressive cases are those where a controlled level of release is too high or, if enforceable, is neglected. Migration from pollution dilution to elimination in many cases is confronted by challenging economical and technological barriers.

Carbon dioxide, while vital for photosynthesis, is sometimes referred to as pollution, because raised levels of the gas in the atmosphere are affecting the Earth's climate. Disruption of the environment can also highlight the connection between areas of pollution that would normally be classified separately, such as those of water and air. Recent studies have investigated the potential for long-term rising levels of atmospheric carbon dioxide to cause slight but critical increases in the acidity of ocean waters, and the possible effects of this on marine ecosystems.

# Chapter 2

# Pollutant

A pollutant is a waste material that pollutes air, water or soil, and is the cause of pollution.

Three factors determine the severity of a pollutant: its chemical nature, its concentration and its persistence. Some pollutants are biodegradable and therefore will not persist in the environment in the long term. However the degradation products of some pollutants are themselves polluting such as the products DDE and DDD produced from degradation of DDT.

## TYPES OF POLLUTANTS

### Stock Pollutants

Pollutants that the environment has little or no absorptive capacity are called stock pollutants (e.g. persistent synthetic chemicals, non-biodegradable plastics, and heavy metals). Stock pollutants accumulate in the environment over time. The damage they cause increases as more pollutant is emitted, and persists as the pollutant accumulates. Stock pollutants can create a burden for future generations by passing on damage that persists well after the benefits received from incurring that damage have been forgotten.

### Fund Pollutants

Fund pollutants are those for which the environment has some absorptive capacity. Fund pollutants do not cause damage to the environment unless the emission rate exceeds the receiving environment's absorptive capacity (e.g. carbon dioxide, which is absorbed by plants and oceans). Fund pollutants are not destroyed, but rather converted into less harmful substances, or diluted/dispersed to non-harmful concentrations.

### Notable Pollutants

Notable pollutants include the following groups:

- Heavy metals
- Persistent organic pollutants
- Polycyclic aromatic hydrocarbons
- Volatile organic compounds
- Environmental xenobiotics

### Zones of Influence

Pollutants can also be defined by their zones of influence, both horizontally and vertically.

### Horizontal Zone

The horizontal zone refers to the area that is damaged by a pollutant. Local pollutants cause damage near the emission source. Regional pollutants cause damage further from the emission source.

### Vertical Zone

The vertical zone is referred to whether the damage is ground-level or atmospheric. Surface pollutants cause damage by concentrations of the pollutant accumulating near the Earth's surface Global pollutants cause damage by concentrations in the atmosphere.

### International

Pollutants can cross international borders and therefore international regulations are needed for their control. The Stockholm Convention on Persistent Organic Pollutants, which entered into force in 2004, is an international legally binding agreement for the control of persistent organic pollutants. Pollutant Release and Transfer Registers (PRTR) are systems to collect and disseminate information on environmental releases and transfers of toxic chemicals from industrial and other facilities.

### European Union

The European Pollutant Emission Register is a type of PRTR providing access to information on the annual emissions of industrial facilities in the Member States of the European Union, as well as Norway.

### United States

***Clean Air Act standards:*** Under the Clean Air Act, the National Ambient Air Quality Standards (NAAQS) are standards developed for outdoor air quality. The National Emissions Standards for Hazardous Air Pollutants are emission standards that are set by the Environmental Protection Agency (EPA) which are not covered by the NAAQS.

***Clean Water Act Standards:*** Under the Clean Water Act, EPA promulgated national standards for municipal sewage treatment plants, also called *publicly owned treatment works,* in the *Secondary Treatment Regulation* National Standards for Industrial Dischargers are called *Effluent Guidelines* (for existing sources) and New Source Performance Standards, and currently cover over 50 industrial categories. In addition, the Act requires states to publish water quality standards for individual water bodies to provide additional protection where the national standards are insufficient.

***RCRA Standards:*** The Resource Conservation and Recovery Act (RCRA) regulates the management, transport and disposal of municipal solid waste, hazardous waste and underground storage tanks.

## Heavy Metal

A heavy metal is a member of a loosely-defined subset of elements that exhibit metallic properties. It mainly includes the transition metals, some metalloids, lanthanides, and actinides. Many different definitions have been proposed—some based on density, some on atomic number or atomic weight, and some on chemical properties or toxicity. The term heavy metal has been called a 'misinterpretation' in an IUPAC technical report due to the contradictory definitions and its lack of a 'coherent scientific basis'. There is an alternative term toxic metal, for which no consensus of exact definition exists either. As discussed below, depending on context, heavy metal can include elements lighter than carbon and can exclude some of the heaviest metals. Heavy metals occur naturally in the ecosystem with large variations in concentration. In modern times, anthropogenic sources of heavy metals, i.e. pollution, have been introduced to the ecosystem. Waste-derived fuels are especially prone to contain heavy metals, so heavy metals are a concern in consideration of waste as fuel.

## Relationship to Living Organisms

Living organisms require varying amounts of 'heavy metals'. Iron, cobalt, copper, manganese, molybdenum, and zinc are required by humans. Excessive levels can be damaging to the organism. Other heavy metals such as mercury, plutonium, and lead are toxic metals that have no known vital or beneficial effect on organisms and their accumulation over time in the bodies of animals can cause serious illness. Certain elements that are normally toxic are, for certain organisms or under certain conditions, beneficial. Examples include vanadium, tungsten, and even cadmium.

## Heavy Metal Pollution

Motivations for controlling heavy metal concentrations in gas streams are diverse. Some of them are dangerous to health or to the environment (e.g. mercury, cadmium, lead, chromium), some may cause corrosion (e.g. zinc, lead), some are harmful in other ways (e.g. arsenic may pollute catalysts). Within the European community the thirteen elements of highest concern are arsenic, cadmium, cobalt, chromium, copper, mercury, manganese, nickel, lead, tin, and thallium, the emissions of which are regulated in waste incinerators. Some of these elements are actually necessary for humans in minute amounts (cobalt, copper, chromium, manganese, nickel) while others are carcinogenic or toxic, affecting, among others, the central nervous system (manganese, mercury, lead, arsenic), the kidneys or liver (mercury, lead, cadmium, copper) or skin, bones, or teeth (nickel, cadmium, copper, chromium).

Heavy metal pollution can arise from many sources but most commonly arises from the purification of metals, e.g., the smelting of copper and the preparation of nuclear fuels. Electroplating is the primary source of chromium and cadmium. Through precipitation of their compounds or by ion exchange into soils and muds, heavy metal pollutants can localize and lay dormant. Unlike organic pollutants, heavy metals do not decay and thus pose a different kind of challenge for remediation. Currently, plants or microrganisms are tentatively used to remove some heavy metals such as mercury. Plants which exhibit hyper accumulation can be used to remove heavy metals from soils by concentrating them in their bio matter. Some treatment of mining tailings has occurred where the vegetation is then incinerated to recover the heavy metals.

One of the largest problems associated with the persistence of heavy metals is the potential for bioaccumulation and biomagnification causing heavier exposure for some organisms than is present in the environment alone. Coastal fish (such as the smooth toadfish) and seabirds (such as the Atlantic Puffin) are often monitored for the presence of such contaminants.

## Medicine

In medical usage, heavy metals are loosely defined and include all toxic metals irrespective of their atomic weight: 'heavy metal poisoning' can possibly include excessive amounts of iron, manganese, aluminium, mercury, cadmium, or beryllium (the fourth lightest element) or such a semimetal as arsenic. This definition excludes bismuth, the heaviest of approximately stable elements, because of its low toxicity.

Minamata disease results from mercury poisoning, and itai-itai disease from cadmium poisoning.

## Hazardous Materials

Heavy metals in a hazardous materials (or 'hazmat') setting are for the most part classified in 'Misc'. on the UN model hazard class, but they are sometimes labeled as a poison when being transported. Nuclear technology Burnup of nuclear fuel is expressed in gigawatt-days per metric ton of heavy metal, where heavy metal means actinides like thorium, uranium, plutonium, etc., including both fissile and fertile material. It does not include elements such as oxygen that may be bonded to the fuel metals, or cladding materials such as zirconium, which might be considered heavy metals by some other standards.

## Persistent Organic Pollutant

Persistent organic pollutants (POPs) are organic compounds that are resistant to environmental degradation through chemical, biological, and photolytic processes. Because of this, they have been observed to persist in the environment, to be capable of long-range transport, bioaccumulate in human and animal tissue, biomagnify in food chains, and to have potential significant impacts on human health and the environment.

Many POPs are currently or were in the past used as pesticides. Others are used in industrial processes and in the production of a range of goods such as solvents, polyvinyl chloride, and pharmaceuticals. Though there are a few natural sources of POPs, most POPs are created by humans in industrial processes, either intentionally or as byproducts.

In May 1995, the United Nations Environment Programme Governing Council (GC) decided to begin investigating POPs, initially beginning with a short list of the following twelve POPs, known as the 'dirty dozen': aldrin, chlordane, DDT, dieldrin, endrin, heptachlor, hexachlorobenzene, mirex, polychlorinated biphenyls, polychlorinated dibenzo-p-dioxins, polychlorinated dibenzofurans, and toxaphene.

Since then, this list has generally been accepted to include such substances as carcinogenic polycyclic aromatic hydrocarbons (PAHs) and certain brominated flame-retardants, as well as some organometallic compounds such as tributyltin (TBT).

The groups of compounds that make up POPs are also classed as PBTs (Persistent, Bioaccumulative and Toxic) or TOMPs (Toxic Organic Micro Pollutants.)

## Chemical Properties

Some of the chemical characteristics of POPs include low water solubility, high lipid solubility, semi-volatility , and high molecular masses. POPs with molecular weights lower than 236 g/mol are less toxic, less persistent in the environment, and have more reversible effects than those with higher molecular masses. POPs are frequently halogenated, usually with chlorine. The more chlorine groups a POP has, the more resistant it is to being broken down over time. One important factor of their chemical properties such as lipid solubility results in the ability to pass through biological phospholipid membranes and bioaccumulate in the fatty tissues of living organisms.

## Long-range Transport

POPs released to the environment have been shown to travel vast distances from their original source. Due to their chemical properties, many POPs are semi-volatile and insoluble. These compounds are therefore unable to transport directly through the environment. The indirect routes include attachment to particulate matter, and through the food chain. The chemicals' semi-volatility allows them to travel long distances through the atmosphere before being deposited. Thus POPs can be found all over the world, including in areas where they have never been used and remote regions such as the middle of oceans and Antarctica. The chemicals' semi-volatility also means that they tend to volatilize in hot regions and accumulate in cold regions, where they tend to condense and stay. PCBs have been found in precipitation.

The ability of POPs to travel great distances is part of the explanation for why countries that banned the use of specific POPs are no longer experiencing a decline in their concentrations; the wind may carry chemicals into the country from places that still use them.

## Health Concerns

POP exposure can cause death and illnesses including disruption of the endocrine, reproductive, and immune systems; neuro-behavioural disorders; and cancers possibly including breast cancer. Exposure to POPs can take place through diet, environmental exposure, or accidents.

A study published in 2006 indicated a link between blood serum levels of POPs and diabetes. Individuals with elevated levels of persistent organic pollutants (DDT, dioxins, PCBs and Chlordane, among others) in their body were found to be up to 38 times more likely to be insulin resistant than individuals with low levels of these pollutants, though the study did not demonstrate a cause and effect relationship. As most exposure to POPs is through consumption of animal fats, study participants with high levels of serum POPs are also very likely to be consumers of high amounts of animal fats, and thus the consumption of the fats themselves, or other associated factors may be responsible for the observed increase in insulin resistance. Another possibility is that insulin resistance causes increased accumulation of POPs. Among study participants, obesity was associated with diabetes only in people who tested high for these pollutants. These pollutants are accumulated in animal fats, so minimizing consumption of animal fats may reduce the risk of diabetes. According to the US Department of Veterans Affairs, type 2 diabetes is on the list of presumptive diseases associated with exposure to Agent Orange (which contained 2, 3, 7, 8 — tetrachlorodibenzodioxin) in the Vietnam War.

## Polycyclic Aromatic Hydrocarbon

Polycyclic aromatic hydrocarbons (PAHs), also known as poly-aromatic hydrocarbons or polynuclear aromatic hydrocarbons are potent atmospheric pollutants that consist of fused aromatic rings and do not contain heteroatoms or carry substituents. Anthracene is the simplest example of a PAH. PAHs occur in oil, coal, and tar deposits, and are produced as byproducts of fuel burning (whether fossil fuel or biomass). As a pollutant, they are of concern because some compounds have been identified as carcinogenic, mutagenic, and teratogenic. PAHs are also found in foods. Studies have shown that high levels of PAHs are found in meat cooked at high temperatures such as grilling or barbecuing, and in smoked fish.

They are also found in the interstellar medium, in comets, and in meteorites and are a candidate molecule to act as a basis for the earliest forms of life. In graphene the PAH motif is extended to large 2D sheets.

Polycyclic aromatic hydrocarbons are lipophilic, meaning they mix more easily with oil than water. The larger compounds are less water-soluble and less volatile (i.e., less prone to evaporate). Because of these properties, PAHs in the environment are found primarily in soil, sediment and oily substances, as opposed to in water or air. However, they are also a component of concern in particulate matter suspended in air.

Natural crude oil and coal deposits contain significant amounts of PAHs, arising from chemical conversion of natural product molecules, such as steroids, to aromatic hydrocarbons. They are also found in processed fossil fuels, tar and various edible oils.

PAHs are one of the most widespread organic pollutants. In addition to their presence in fossil fuels they are also formed by incomplete combustion of carbon-containing fuels such as wood, coal, diesel, fat, tobacco, and incense. Different types of combustion yield different distributions of PAHs in both relative amounts of individual PAHs and in which isomers are produced. Thus, coal burning produces a different mixture than motor-fuel combustion or a forest fire, making the compounds potentially useful as indicators of the burning history. Hydrocarbon emissions from fossil fuel-burning engines are regulated in developed countries.

PAHs toxicity is very structurally dependent, with isomers (PAHs with the same formula and number of rings) varying from being non-toxic to being extremely toxic. Thus, highly carcinogenic PAHs may be small or large. One PAH compound, benzo[*a*]pyrene, is notable for being the first chemical carcinogen to be discovered (and is one of many carcinogens found in cigarette smoke). The EPA has classified seven PAH compounds as probable human carcinogens: benzo[*a*]anthracene, benzo[*a*]pyrene, benzo[*b*]fluoranthene, benzo[*k*]fluoranthene, chrysene, dibenz(ah) anthracene, and indeno(1, 2, 3-cd) pyrene.

PAHs known for their carcinogenic, mutagenic and teratogenic properties are benzo[*a*]anthracene and chrysene, benzo[*b*]fluoranthene, benzo[*j*]fluoranthene, benzo[*k*]fluoranthene, benzo[*a*]pyrene, benzo[*ghi*]perylene, coronene, dibenz(ah)anthracene ($C_{20}H_{14}$), indeno(1,2,3-cd)pyrene ($C_{22}H_{12}$) and ovalene.

High prenatal exposure to PAH is associated with lower IQ and childhood asthma. The Center for Children's Environmental Health reports studies that demonstrate that exposure to PAH pollution during pregnancy is related to adverse birth outcomes including low birth weight, premature delivery, and heart malformations. Cord blood of exposed babies shows DNA damage that has been linked to cancer. Follow-up studies show a higher level of developmental delays at age three, lower scores on IQ tests and increased behavorial problems at ages six and eight.

## Chemistry

The simplest PAHs, as defined by the International Union of Pure and Applied Chemistry (IUPAC) (G.P Moss, IUPAC nomenclature for fused-ring systems), are phenanthrene and anthracene, which both contain three fused aromatic rings. Smaller molecules, such as benzene, are not PAHs.

PAHs may contain four-, five-, six- or seven-member rings, but those with five or six are most common. PAHs composed only of six-membered rings are called alternant PAHs. Certain alternant PAHs are called benzenoid PAHs. The name comes from benzene, an aromatic hydrocarbon with a single, six-membered ring. These can be benzene rings interconnected with each other by single carbon-carbon bonds and with no rings remaining that do not contain a complete benzene ring.

The set of alternant PAHs is closely related to a set of mathematical entities called polyhexes, which are planar figures composed by conjoining regular hexagons of identical size.

PAHs containing up to six fused aromatic rings are often known as 'small' PAHs, and those containing more than six aromatic rings are called 'large' PAHs. Due to the availability of samples of the various small PAHs, the bulk of research on PAHs has been of those of up to six rings. The biological activity and occurrence of the large PAHs does appear to be a continuation of the small PAHs. They are found as combustion products, but at lower levels than the small PAHs due to the kinetic limitation of their production through addition of successive rings. In addition, with many more isomers possible for larger PAHs, the occurrence of specific structures is much smaller.

PAHs possess very characteristic UV absorbance spectra. These often possess many absorbance bands and are unique for each ring structure. Thus, for a set of isomers, each isomer has a different UV absorbance spectrum than the others. This is particularly useful in the identification of PAHs. Most PAHs are also fluorescent, emitting characteristic wavelengths of light when they are excited (when the molecules absorb light). The extended pi-electron electronic structures of PAHs lead to these spectra, as well as to certain large PAHs also exhibiting semi-conducting and other behaviours.

Naphthalene ($C_{10}H_8$ constituent of mothballs), consisting of two coplanar six-membered rings sharing an edge, is another aromatic hydrocarbon. By formal convention, it is not a true PAH, though is referred to as a bicyclic aromatic hydrocarbon.

Aqueous solubility decreases approximately one order of magnitude for each additional ring.

In January 2004 (at the 203rd Meeting of the American Astronomical Society), it was reported that a team led by A. Witt of the University of Toledo, Ohio studied ultraviolet light emitted by the Red Rectangle nebula and found the spectral signatures of anthracene and pyrene (no other such complex molecules had ever before been found in space). This discovery was considered as a controversial confirmation of a hypothesis that as nebulae of the same type as the Red Rectangle approach the ends of their lives, convection currents cause carbon and hydrogen in the nebulae's core to get caught in stellar winds, and radiate outward. As they cool, the atoms supposedly bond to each other in various ways and eventually form particles of a million or more atoms. Witt and his team inferred that since they discovered PAHs—which may have been vital in the formation of early life on Earth—in a nebula, by necessity they must originate in nebulae. More recently, PAHs, along with fullerenes (or 'buckyballs'), have been detected in other nebulae. Fullerenes are also implicated in the origin of life; according to astronomer Letizia Stanghellini, "It's possible that buckyballs from outer space provided seeds for life on Earth".

## Volatile Organic Compound

Volatile organic compounds (VOCs) refers to organic chemical compounds which have significant vapor pressures and which can affect the environment and human health. VOCs are numerous, varied, and ubiquitous. Although VOCs include both man-made and naturally occurring chemical compounds, it is the anthropogenic VOCs that are regulated, especially for indoors where concentrations can be highest. VOCs are typically not acutely toxic but have chronic effects. Because the concentrations are usually low and the symptoms slow to develop, analysis of VOCs and their effects is a demanding area.

The definitions of VOCs used for control of precursors of photochemical smog used by EPA and states with their own outdoor air pollution regulations includes exemptions for compounds that are technically only those volatile organic compounds but that are determined to be non-reactive or of low-reactivity in the smog formation process. EPA formerly defined these compounds as Reactive Organic Gases (ROG) but changed the terminology to VOC for simplicity's sake.

In the USA, different regulation exists per state — most prominent is the VOC regulation by SCAQMD and by California ARB. However, this specific use of the term VOCs can be misleading, specifically when applied to indoor air quality because many chemicals that are not regulated for controlling outdoor air pollution still can be important for indoor air pollution — there is no correlation between VOC content in a product and VOC emissions from that product into indoor air.

### Canada

Health Canada classes VOCs as organic compounds that have boiling points roughly in the range of 50 to 250 °C (122 to 482 °F). The emphasis is placed on commonly encountered VOCs which would have an effect on air quality.

### European Union

A VOC is any organic compound having an initial boiling point less than or equal to 250° C measured at a standard atmospheric pressure of 101.3 kPa and can do damage to visual or audible senses.

### US

VOCs (or specific subsets of the VOCs) are legally defined in the various laws and codes under which they are regulated. Other definitions may be found from government agencies investigating or advising about VOCs. The United States Environmental Protection Agency (EPA) regulates VOCs in the air, water, and land. The Safe Drinking Water Act implementation includes a list labeled "VOCs in connection with contaminants which are organic and volatile." The EPA also publishes testing methods for chemical compounds, some of which refer to VOCs.

In addition to drinking water, VOCs are regulated in discharges to waters (sewage treatment and stormwater disposal), as hazardous waste, but not in non industrial indoor air. The United States Department of Labour and its Occupational Safety and Health Administration (OSHA) regulate VOC exposure in the workplace. Volatile organic compounds which are hazardous material would be regulated by the Pipeline and Hazardous Materials Safety Administration while being transported.

### Biologically Derived VOCs

The majority of VOCs arise from plants. One indication of this flux is the strong odour emitted by many plants. The emissions are affected by a variety of factors, such as temperature, which determines rates of volatilization and growth, and sunlight, which determines rates of biosynthesis. Emission occurs almost exclusively from the leaves, the stomata in particular. A major class of VOCs are terpenes, such as myrcene. Providing a sense of scale, a forest 62,000 $km^2$ in area (the U.S. state of Pennsylvania) is estimated to emit 3,400,000 kilograms of terpenes on a typical August day during the growing season. Induction of genes producing volatile organic compounds and subsequent increase in volatile terpenes has been achieved in maize using (Z)-3-Hexen-1-ol and other plant hormones.

## MAJOR ANTHROPOGENIC SOURCES

### Paints and Coatings

A major source of man-made VOCs are solvents, especially paints and protective coatings. Solvents are required to spread a protective or decorative film. Approximately 12 billion litres of paints are produced annually. Typical solvents are aliphatic hydrocarbons, ethyl acetate, glycol ethers, and acetone. Motivated by cost, environmental concerns, and regulation, the paint and coating industries are increasingly shifting towards aqueous solvents.

### Chlorofluorocarbons and Chlorocarbons

Chlorofluorocarbons, which are banned or highly regulated, were widely used cleaning products and refrigerants. Tetrachloroethene is used widely in dry cleaning and by industry. Industrial use of fossil fuels produces VOCs either directly as products (e.g. gasoline) or indirectly as byproducts (e.g. automobile exhaust).

## MTBE

MTBE was banned in the US around 2004 in order to limit further contamination of drinking water aquifers.

## Indoor Air

Since people today spend most of their time at home or in an office, long-term exposure to VOCs in the indoor environment can contribute to sick building syndrome. In offices, VOC results from new furnishings, wall coverings, and office equipment such as photocopy machines, which can off-gas VOCs into the air. Good ventilation and air conditioning systems are helpful at reducing VOC emissions in the indoor environment. Studies also show that relative leukemia and lymphoma can increase through prolonged exposure of VOCs in the indoor environment.

The United States Environmental Protection Agency (EPA) has found concentrations of VOCs in indoor air commonly to be 2 to 5 times greater than in outdoor air and sometimes far greater. During certain activities indoor levels of VOCs may reach 1,000 times that of the outside air. Studies have shown that individual VOC emissions by themselves are not that high in an indoor environment, but the indoor total VOC (TVOC) concentrations can be up to five times higher than the VOC outdoor levels. New buildings especially, contribute to the highest level of VOC off-gassing in an indoor environment because of the abundant new materials generating VOC particles at the same time in such a short time period. In addition to new buildings, we also use many consumer products that emit VOC compounds, therefore the total concentration of VOC levels is much greater within the indoor environment.

VOC concentration in an indoor environment is three to four times higher than the VOC concentrations during the summer. High indoor VOC levels are attributed to the low rates of air exchange between the indoor and outdoor environment as a result of tight-shut windows and the increasing use of humidifiers.

## Indoor Air Quality and Emissions into Indoor Air

In most countries a separate definition of VOCs is used with regard to indoor air quality that comprises each organic chemical compound that can be measured as follows: Adsorption from air on Tenax TA, thermal desorption, gas chromatographic separation over a 100 per cent non polar column (dimethylpolysiloxane). VOC (volatile organic compounds) are all compounds that appear in the gas chromatogram between and including n-hexane and n-hexadecane. Compounds appearing earlier are called VVOC (very volatile organic compounds) compounds appearing later are called

SVOC (semi-volatile organic compounds). See also these standards: ISO 16000-6, ISO 13999-2, VDI 4300-6, German AgBB evaluating scheme, German DIBt approval scheme, GEV testing method for the EMICODE. Some overviews over VOC emissions rating schemes have been collected and compared.

**Formaldehyde**

Many building materials such as paints, adhesives, wall boards, and ceiling tiles slowly emit formaldehyde, which irritates the mucous membranes and can make a person irritated and uncomfortable. Formaldehyde emissions from wood are in the range of 0.02 - 0.04 ppm. Relative humidity within an indoor environment can also affect the emissions of formaldehyde. High relative humidity and high temperatures allow more vapourization of formaldehyde from wood-materials. There are also many sources of VOCs in office buildings, which include new furnishings, wall coverings, and office equipment such as photocopy machines which can off-gas VOCs into the air.

**Health Risks**

Respiratory, allergic, or immune effects in infants or children are associated with man-made VOCs and other indoor or outdoor air pollutants.

Some VOCs, such as styrene and limonene, can react with nitrogen oxides or with ozone to produce new oxidation products and secondary aerosols, which can cause sensory irritation symptoms. Unspecified VOCs are important in the creation of smog.

Health effects include:

Eye, nose, and throat irritation; headaches, loss of coordination, nausea; damage to liver, kidney, and central nervous system. Some organics can cause cancer in animals; some are suspected or known to cause cancer in humans. Key signs or symptoms associated with exposure to VOCs include conjunctival irritation, nose and throat discomfort, headache, allergic skin reaction, dyspnea, declines in serum cholinesterase levels, nausea, emesis, epistaxis, fatigue, dizziness.

The ability of organic chemicals to cause health effects varies greatly from those that are highly toxic, to those with no known health effect. As with other pollutants, the extent and nature of the health effect will depend on many factors including level of exposure and length of time exposed. Eye and respiratory tract irritation, headaches, dizziness, visual disorders, and memory impairment are among the immediate symptoms that some people have experienced soon after exposure to some organics. At present, not much is known about what health effects occur from the levels of organics usually found in homes. Many organic compounds are known to cause cancer in animals; some are suspected of causing, or are known to cause, cancer in humans.

## Limit Values for VOC Emissions

Limit values for VOC emissions into indoor air are published by e.g. AgBB, AFSSET, California Department of Public Health, and others.

## Chemical Fingerprinting

The exhaled human breath contains a few hundred volatile organic compounds and is used in breath analysis to serve as a VOC biomarker to test for diseases such as lung cancer. One study has shown that "volatile organic compounds . . . are mainly blood borne and therefore enable monitoring of different processes in the body." And it appears that VOC compounds in the body "may be either produced by metabolic processes or inhaled/absorbed from exogenous sources" such as environmental tobacco smoke. Research is still in the process to determine whether VOCs in the body are contributed by cellular processes or by the cancerous tumours in the lung or other organs.

VOCs in the environment or certain atmospheres can be detected based in different principles and interactions between the organic compounds and the sensor components. There are electronic devices that can detect ppm concentrations despite the non-selectivity. Others can predict with reasonable accuracy the molecular structure of the volatile organic compounds in the environment or enclosed atmospheres and could be used as accurate monitors of the Chemical Fingerprint and further as health monitoring devices.

## Environmental Xenobiotic

*Environmental* xenobiotic are xenobiotic substances with a biological activity that are found as pollutants in the natural environment.

## Pharmaceuticals

Pharmaceutical drug is a chemical used for the alteration, diagnosis, prevention and treatment of disease, health conditions or structure/function of the human body. Pharmaceutically active compounds (PhACs) are those pharmaceuticals that have by one route or another entered the environment as the parent compound or as pharmacologically active metabolites. Drugs are developed with the intention of having a beneficial biological effect on the organism to which they are administered, but many such compounds all too often pass into the environment where they may exert an unwanted biological effect.

For many years PhACs have been all but ignored as environmental researchers concentrated on the well known environmentally dangerous chemicals that were/are largely used in agriculture and industry. But with increasing technology to help in the separation and identification of multiple compounds in a mixture, PhACs and their effects have received increasing attention. PhACs have not (until relatively recently) been seen as potentially toxic because regulations associated with pharmaceuticals are typically overseen by human health organizations which have limited experience with environmental issues.

Nearly all categories of pharmaceuticals including pain killers (analgesics and anti-inflammatory), antibiotics (antibacterial), anticonvulsant drugs, Beta-blockers, blood lipid regulators, X-ray contrast media, cytostatic drugs (Chemotherapy), oral contraceptives, and veterinary pharmaceuticals among many others have been found in the environment.

**Sources and Origins**

PhACs can be entered into the environment in two main ways; direct and indirect. Indirect sources are PhACs that have performed their biologically intended effect and are passed onto the environment in either their complete or a modified state.

PhAC's can be discharged directly by manufacturers of the pharmaceuticals or effluents from hospitals. However with increasing regulation by local, state and federal regulating agencies, direct discharge is becoming much less of an issue.

There are also several indirect sources of PhACs into the environment. One common indirect source of PhACs into the environment is the passing of antibiotics, anesthetics and growth promoting hormones by domesticated animals in urine and manure. This is often stored in large pits before being pumped and applied to fields as fertilizers where many of the PhACs can be washed away by rainfall to aquatic environments.

Family pets can also be an indirect source of PhACs into the environment.

Most of the PhACs in the environment however come from human sources. A direct human source is leachate from a landfill. Often the pharmaceuticals that are located in landfills are found in their original, most chemically active state.

Most pharmaceuticals are administered and passed through the human body in one of three ways:

1. Metabolized partially or completely within the body and made inactive (Ideal).

2. Partially metabolized and passed through the system.
3. Passed through the body unmodified (Worst Case Scenario). In any manner PhACs are then passed to sewage treatment plants (STPs), where facilities are designed to break down natural human waste by microbial degradation. However, many PhACs are of very complex structure and are incompletely broken down in STPs before they are passed into the environment.

## Fate in Environment

Once PhACs are entered into the environment they suffer one of three fates:

1. Biodegradation into carbon dioxide and water.
2. Undergo some form of degradation and form metabolites.
3. Persist in the environment unmodified. The amount of the compound that is broken down depends on several factors such as bioavailability and compound structure among many others.

## Effects

Because PhACs have come into the limelight relatively recently their effects on the environment are not completely understood. PhACs are also not generally intended to come in contact with the environment, and therefore are not typically tested environmentally prior to release. Therefore several tests are required to determine the different mechanisms and side effects of PhACs in the environment making testing largely impractical.

Many PhACs have very broad modes of action in humans. Similar, subtle reactions may occur in organisms in the environment that are not easily seen by humans. Highly specific mechanisms in humans may solicit profound effects at extremely low concentrations. Many effects may not necessarily be readily detectable and lead to ecological change that would be erroneously attributed to natural change. This said there are several effects that have been identified in the literature.

One long term, possibly irreversible effect is microbiological resistance to antibiotics (antibiotic resistance). Some bacteria may be able to survive when administered antibiotics (especially at low concentrations) . Those colonies will multiply and produce new colonies that are resistant to that particular antibiotic and will not succumb the next time antibiotics are administered. Because rivers and streams are ever flowing objects they are an ideal pathway for antibiotics to reach bacteria and therefore provide a source and reservoir for resistant strains to develop and establish themselves.

Another recent discovery is endocrine disruptors. Endocrine disruptors can replace or disturb the balance of hormones within an organism and have been found to be occurring in waters with a concentration in the ng/L level for certain compounds. Some possible effects of endocrine disruptors are male and female sterility, feminization of males, masculinization of females and abnormal testes growth among many others. The exact pathway of occurrence of endocrine disruptors is not completely certain, however several pathways have been proposed.

Typically PhACs are found in low concentrations, (<1 μg/L) making acute toxicity effects fairly unlikely. However, because of their continual input to the environment it is possible for chronic toxicity effects to occur. One major area of concern with several compounds being present at low levels at the same time is what happens when the compounds mix? It is possible and truly likely that these mixtures will have additive, neutralistic or synergistic effects. But again testing would be both time consuming and very expensive to test all of the combined effects.

CHAPTER 3

# DDT [Dichlorodiphenyltrichloroethane]

DDT (from its trivial name, dichlorodiphenyltrichloroethane) is one of the most well-known synthetic pesticides.

First synthesized in 1874, DDT's insecticidal properties were not discovered until 1939, and it was used with great success in the second half of World War II to control malaria and typhus among civilians and troops. The Swiss chemist Paul Hermann Müller was awarded the Nobel Prize in Physiology or Medicine in 1948 "for his discovery of the high efficiency of DDT as a contact poison against several arthropods." After the war, DDT was used as an agricultural insecticide, and soon its production and use skyrocketed.

In 1962, American biologist Rachel Carson wrote *Silent Spring.* The book cataloged the environmental impacts of indiscriminate DDT use in the US and questioned the logic of releasing large amounts of chemicals into the environment without fully understanding their effects on the environment or human health. The book suggested that DDT and other pesticides cause cancer and that their agricultural use was a threat to wildlife, particularly birds. Its publication was a signature event in the birth of the environmental movement. It produced a large public outcry that led to a 1972 ban in the US. DDT was subsequently banned for agricultural use worldwide under the Stockholm Convention, but limited, controversial use in disease vector control continues.

Along with the Endangered Species Act, the US DDT ban is cited by scientists as a major factor in the comeback of the bald eagle, the national bird of the United States, from near-extinction in the contiguous US.

DDT is an organochlorine, similar in structure to the insecticide methoxychlor and the acaricide dicofol. It is a highly hydrophobic, colourless, crystalline solid with a weak, chemical odour. It is nearly insoluble in water but has a good solubility in most organic solvents, fats, and oils. DDT does not occur naturally, but is produced by the reaction of chloral ($CCl_3CHO$) with chlorobenzene ($C_6H_5Cl$) in the presence of sulfuric acid, which acts as a catalyst. Trade names that DDT has been marketed under include Anofex, Cezarex, Chlorophenothane, Clofenotane, Dicophane, Dinocide, Gesarol, Guesapon, Guesarol, Gyron, Ixodex, Neocid, Neocidol, and Zerdane.

## Isomers and Related Compounds

Commercial DDT is a mixture of several closely-related compounds. The major component (77%) is the *p,p'* isomer. The *o, p'* isomer is also present in significant amounts (15%). Dichlorodiphenyldichloroethylene (DDE) and dichlorodiphenyldichloroethane (DDD) make up the balance. DDE and DDD are also the major metabolites and breakdown products in the environment. The term 'total DDT' is often used to refer to the sum of all DDT related compounds (*p,p'*-DDT, *o,p'*-DDT, DDE, and DDD) in a sample.

## Production and Use Statistics

From 1950 to 1980, DDT was extensively used in agriculture — more than 40,000 tonnes were used each year worldwide — and it has been estimated that a total of 1.8 million tonnes have been produced globally since the 1940s. In the U.S., where it was manufactured by Ciba, Montrose Chemical Company, Pennwalt and Velsicol Chemical Corporation, production peaked in 1963 at 82,000 tonnes per year. More than 600,000 tonnes (1.35 billion lbs) were applied in the U.S. before the 1972 ban. Usage peaked in 1959 at about 36,000 tonnes.

Today, 4-5,000 tonnes are used globally each year for the control of malaria and visceral leishmaniasis. India is the largest consumer. India, China, and North Korea are the only countries that still produce and export. Production is reportedly rising.

In insects it opens sodium ion channels in neurons, causing them to fire spontaneously, which leads to spasms and eventual death. Insects with certain mutations in their sodium channel gene are resistant to DDT and other similar insecticides. DDT resistance is also conferred by up-regulation of genes expressing cytochrome P450 in some insect species.

In humans, however, it may affect health through genotoxicity or endocrine disruption.

First synthesized in 1874 by Othmar Zeidler, DDT's insecticidal properties were not discovered until 1939 by the Swiss scientist Paul Hermann Müller, who was awarded the 1948 Nobel Prize in Physiology and Medicine for his efforts.

## Use in the 1940s and 1950s

DDT is the best-known of several chlorine-containing pesticides used in the 1940s and 1950s. With pyrethrum in short supply, DDT was used extensively during World War II by the Allies to control the insect vectors of typhus — nearly eliminating the disease in many parts of Europe. In the South Pacific, it was sprayed aerially for malaria control with spectacular effects. While DDT's chemical and insecticidal properties were important factors in these victories, advances in application equipment coupled with a high degree of organization and sufficient manpower were also crucial to the success of these programmes. In 1945, it was made available to farmers as an agricultural insecticide, and it played a minor role in the final elimination of malaria in Europe and North America. By the time DDT was introduced in the U.S., the disease had already been brought under control by a variety of other means. One CDC physician involved in the United States' DDT spraying campaign said of the effort that "we kicked a dying dog".

In 1955, the World Health Organization commenced a programme to eradicate malaria worldwide, relying largely on DDT. The programme was initially highly successful, eliminating the disease in "Taiwan, much of the Caribbean, the Balkans, parts of northern Africa, the northern region of Australia, and a large swath of the South Pacific" and dramatically reducing mortality in Sri Lanka and India. However widespread agricultural use led to resistant insect populations. In many areas, early victories partially or completely reversed, and in some cases rates of transmission even increased. The programme was successful in eliminating malaria only in areas with "high socio-economic status, well-organized healthcare systems, and relatively less intensive or seasonal malaria transmission".

DDT was less effective in tropical regions due to the continuous life cycle of mosquitoes and poor infrastructure. It was not applied at all in sub-Saharan Africa due to these perceived difficulties. Mortality rates in that area never declined to the same dramatic extent, and now constitute the bulk of malarial deaths worldwide, especially following the disease's resurgence as a result of resistance to drug treatments and the spread of the deadly malarial variant caused by *Plasmodium falciparum*. The goal of eradication was abandoned in 1969, and attention was focussed on controlling and treating the disease. Spraying programmes (especially using DDT) were curtailed due to concerns over safety and environmental effects, as well as problems in administrative, managerial and financial implementation, but mostly because mosquitoes were developing resistance to DDT. Efforts shifted from spraying to the use of bednets impregnated with insecticides and other interventions.

As early as the 1940s, scientists in the U.S. had begun expressing concern over possible hazards associated with DDT, and in the 1950s the government began tightening some of the regulations governing its use. However, these early events received little attention, and it was not until 1957, when the *New York Times* reported an unsuccessful struggle to restrict DDT use in Nassau County, New York, that the issue came to the attention of the popular naturalist-author, Rachel Carson. William Shawn, editor of *The New Yorker*, urged her to write a piece on the subject, which developed into her famous book *Silent Spring*, published in 1962. The book argued that pesticides, including DDT, were poisoning both wildlife and the environment and were also endangering human health.

Silent Spring was a best seller, and public reaction to it launched the modern environmental movement in the United States. The year after it appeared, President Kennedy ordered his Science Advisory Committee to investigate Carson's claims. The report the committee issued "add[ed] up to a fairly thorough-going vindication of Rachel Carson's Silent Spring thesis," in the words of the journal *Science*, and recommended a phaseout of 'persistent toxic pesticides'. DDT became a prime target of the growing anti-chemical and anti-pesticide movements, and in 1967 a group of scientists and lawyers founded the Environmental Defense Fund (EDF) with the specific goal of winning a ban on DDT. Victor Yannacone, Charles Wurster, Art Cooley and others associated with inception of EDF had all witnessed bird kills or declines in bird populations and suspected that DDT was the cause. In their campaign against the chemical, EDF petitioned the government for a ban and filed a series of lawsuits. Around this time, toxicologist David Peakall was measuring DDE levels in the eggs of peregrine falcons and California condors and finding that increased levels corresponded with thinner shells.

In response to an EDF suit, the U.S. District Court of Appeals in 1971 ordered the EPA to begin the de-registration procedure for DDT. After an initial six-month review process, William Ruckelshaus, the Agency's first Administrator rejected an immediate suspension of DDT's registration, citing studies from the EPA's internal staff stating that DDT was not an imminent danger to human health and wildlife. However, the findings of these staff members were criticized, as they were performed mostly by economic entomologists inherited from the United States Department of Agriculture, whom many environmentalists felt were biased towards agribusiness and tended to minimize concerns about human health and wildlife. The decision not to ban thus created public controversy.

The EPA then held seven months of hearings in 1971-1972, with scientists giving evidence both for and against the use of DDT. In the summer of 1972, Ruckelshaus announced the cancellation of most uses of DDT—an exemption allowed for public health uses under some conditions. Immediately after the cancellation was announced, both EDF and the DDT manufacturers filed suit against the EPA, with the industry seeking to overturn the ban, and EDF seeking a comprehensive ban. The cases were consolidated, and in 1973 the U.S. Court of Appeals for the District of Columbia ruled that the EPA had acted properly in banning DDT.

The U.S. DDT ban took place amidst a growing public mistrust of industry, with the Surgeon General issuing a report on smoking in 1964, the Cuyahoga River catching fire in 1969, the fiasco surrounding the use of diethylstilbestrol (DES), and the well-publicized decline in the bald eagle population.

Some uses of DDT continued under the public health exemption. For example, in June 1979, the California Department of Health Services was permitted to use DDT to suppress flea vectors of bubonic plague. DDT also continued to be produced in the US for foreign markets until as late as 1985, when over 300 tonnes were exported.

**Restrictions on Usage**

In the 1970s and 1980s, agricultural use was banned in most developed countries, beginning with Hungary in 1968 then in Norway and Sweden in 1970, Germany and the US in 1972, but not in the United Kingdom until 1984. Vector control use has not been banned, but it has been largely replaced by less persistent alternative insecticides.

The Stockholm Convention, which took effect in 2004, outlawed several persistent organic pollutants, and restricted DDT use to vector control. The Convention has been ratified by more than 160 countries and is endorsed by most environmental groups. Recognizing that total elimination in many malaria-prone countries is currently unfeasible because there are few affordable or effective alternatives, public health use is exempt from the ban pending acceptable alternatives. Malaria Foundation International states, "The outcome of the treaty is arguably better than the status quo going into the negotiations . . . For the first time, there is now an insecticide which is restricted to vector control only, meaning that the selection of resistant mosquitoes will be slower than before"

Despite the worldwide ban, agricultural use continues in India North Korea, and possibly elsewhere.

Today, about 4-5,000 tonnes each year are used for vector control. DDT is applied to the inside walls of homes to kill or repel mosquitoes. This intervention, called indoor residual spraying (IRS), greatly reduces environmental damage. It also reduces the incidence of DDT resistance.

For comparison, treating 40 hectares (99 acres) of cotton during a typical U.S. growing season requires the same amount of chemical as roughly 1,700 homes.

DDT is a persistent organic pollutant that is extremely hydrophobic and strongly absorbed by soil. Depending on conditions, its soil half life can range from 22 days to 30 years. Routes of loss and degradation include runoff, volatilization, photolysis and aerobic and anaerobic biodegradation. When applied to aquatic ecosystems it is quickly absorbed by organisms and by soil or it evaporates, leaving little DDT dissolved in the water itself. Its breakdown products and metabolites, DDE and DDD, are also highly persistent and have similar chemical and physical properties. DDT and its breakdown products are transported from warmer regions of the world to the Arctic by the phenomenon of global distillation, where they then accumulate in the region's food web.

DDT, DDE, and DDD magnify through the food chain, with apex predators such as raptor birds concentrating more chemicals than other animals in the same environment. They are very lipophilic and are stored mainly in body fat. DDT and DDE are very resistant to metabolism; in humans, their half-lives are 6 and up to 10 years, respectively. In the United States, these chemicals were detected in almost all human blood samples tested by the Centers for Disease Control in 2005, though their levels have sharply declined since most uses were banned in the US. Estimated dietary intake has also declined, although FDA food tests commonly detect it.

Marine macroalgae (seaweed) help reduce soil toxicity by up to 80 per cent within six weeks.

**Effects on Wildlife and Eggshell Thinning**

DDT is toxic to a wide range of animals in addition to insects, including marine animals such as crayfish, daphnids, sea shrimp and many species of fish. It is less toxic to mammals, but may be moderately toxic to some amphibian species, especially in the larval stage. Most famously, it is a reproductive toxicant for certain birds species, and it is a major reason for the decline of the bald eagle, brown pelican peregrine falcon, and osprey. Birds of prey, waterfowl, and song birds are more susceptible to eggshell thinning than chickens and related species, and DDE appears to be more potent than DDT. Even in 2010, more than forty years after the U.S. ban, California condors which feed on sea lions at Big Sur which in turn feed in the Palos Verdes Shelf area of the Montrose Chemical Superfund site seemed to be having continued thin-shell problems. Scientists with the Ventana Wildlife Society and others are intensifying studies and remediations of the condors' problems.

The biological thinning mechanism is not entirely known, but there is strong evidence that p,p'-DDE inhibits calcium ATPase in the membrane of the shell gland and reduces the transport of calcium carbonate from blood into the eggshell gland. This results in a dose-dependent thickness reduction. There is also evidence that o,p'-DDT disrupts female reproductive tract development, impairing eggshell quality later. Multiple mechanisms may be at work, or different mechanisms may operate in different species. Some studies show that although DDE levels have fallen dramatically, eggshell thickness remains 10-12 per cent thinner than before DDT was first used.

## Effects on Human Health

Potential mechanisms of action on humans are genotoxicity and endocrine disruption. DDT may be directly genotoxic, but may also induce enzymes to produce other genotoxic intermediates and DNA adducts. It is an endocrine disruptor; The DDT metabolite DDE acts as an antiandrogen (but not as an estrogen). p,p'-DDT, DDT's main component, has little or no androgenic or estrogenic activity. Minor component o,p'-DDT has weak estrogenic activity.

## Acute Toxicity

DDT is classified as 'moderately toxic' by the United States National Toxicology Programme (NTP) and 'moderately hazardous' by the World Health Organization (WHO), based on the rat oral $LD_{50}$ of 113 mg/kg. DDT has on rare occasions been administered orally as a treatment for barbiturate poisoning.

## Chronic Toxicity

### *Diabetes*

Organochlorine compounds, generally, and DDT and DDE, specifically, have been linked to diabetes. A number of studies from the US, Canada, and Sweden have found that the prevalence of the disease in a population increases with serum DDT or DDE levels.

## Developmental and Reproductive Toxicity

DDT and DDE, like other organochlorines, have been shown to have xenoestrogenic activity, meaning they are chemically similar enough to estrogens to trigger hormonal responses in animals. This endocrine disrupting activity has been observed in mice and rat toxicological studies, and available epidemiological evidence indicates that these effects may be occurring in humans as a result of DDT exposure. These effects may cause developmental and reproductive toxicity:

- A review article in *The Lancet* states, "research has shown that exposure to DDT at amounts that would be needed in malaria control might cause preterm birth and early weaning ... toxicological evidence shows endocrine-disrupting properties; human data also indicate possible disruption in semen quality, menstruation, gestational length, and duration of lactation".
- Human epidemiological studies suggest that exposure is a risk factor for premature birth and low birth weight, and may harm a mother's ability to breast feed Some 21st century researchers argue that these effects may increase infant deaths, offsetting any anti-malarial benefits.

A 2008 study, however, failed to confirm the association between exposure and difficulty breastfeeding.

- Several recent studies demonstrate a link between *in utero* exposure to DDT or DDE and developmental neurotoxicity in humans. For example, a 2006 University of California, Berkeley study suggests that children exposed while in the womb have a greater chance of development problems, and other studies have found that even low levels of DDT or DDE in umbilical cord serum at birth are associated with decreased attention at infancy and decreased cognitive skills at 4 years of age Similarly, Mexican researchers have linked first trimester DDE exposure to retarded psychomotor development.
- Other studies document decreases in semen quality among men with high exposures (generally from IRS).
- Studies generally find that high blood DDT or DDE levels do not increase time to pregnancy (TTP). There is some evidence that the daughters of highly exposed women may have more difficulty getting pregnant (i.e. increased TTP).
- DDT is associated with early pregnancy loss, a type of miscarriage. A prospective cohort study of Chinese textile workers found "a positive, monotonic, exposure-response association between preconception serum total DDT and the risk of subsequent early pregnancy losses." The median serum DDE level of study group was lower than that typically observed in women living in homes sprayed with DDT.
- A Japanese study of congenital hypothyroidism concluded that *in utero* DDT exposure may affect thyroid hormone levels and "play an important role in the incidence and/or causation of cretinism." Other studies have also found the DDT or DDE interfere with proper thyroid function.

### Other

Occupational exposure in agriculture and malaria control has been linked to neurological problems (i.e. Parkinsons) and asthma.

### Use Against Malaria

Malaria remains a major public health challenge in many countries. 2008 WHO estimates were 243 million cases, and 863,000 deaths. About 89 per cent of these deaths occur in Africa, and mostly to children under the age of 5. DDT is one of many tools that public health officials use to fight the disease. Its use in this context has been called everything from a "miracle weapon [that is] like Kryptonite to the mosquitoes," to 'toxic colonialism'.

Before DDT, eliminating mosquito breeding grounds by drainage or poisoning with Paris green or pyrethrum was sometimes successful in fighting malaria. In parts of the world with rising living standards, the elimination of malaria was often a collateral benefit of the introduction of window screens and improved sanitation. Today, a variety of usually simultaneous interventions is the norm. These include antimalarial drugs to prevent or treat infection; improvements in public health infrastructure to quickly diagnose, sequester, and treat infected individuals; bednets and other methods intended to keep mosquitoes from biting humans; and vector control strategies such as larvaciding with insecticides, ecological controls such as draining mosquito breeding grounds or introducing fish to eat larvae, and indoor residual spraying with insecticides, possibly including DDT. IRS involves the treatment of all interior walls and ceilings with insecticides, and is particularly effective against mosquitoes, since many species rest on an indoor wall before or after feeding. DDT is one of 12 WHO—approved IRS insecticides. How much of a role DDT should play in this mix of strategies is still controversial.

WHO's anti-malaria campaign of the 1950s and 1960s relied heavily on DDT and the results were promising, though temporary. Experts tie the resurgence of malaria to multiple factors, including poor leadership, management and funding of malaria control programmes; poverty; civil unrest; and increased irrigation. The evolution of resistance to first-generation drugs (e.g. chloroquine) and to insecticides exacerbated the situation. Resistance was largely fueled by often unrestricted agricultural use. Resistance and the harm both to humans and the environment led many governments to restrict or curtail the use of DDT in vector control as well as agriculture.

Once the mainstay of anti-malaria campaigns, as of 2008 only 12 countries used DDT, including India and some southern African states, though the number is expected to rise.

### Carcinogenicity

DDT is suspected to cause cancer. The NTP classifies it as "reasonably anticipated to be a carcinogen," the International Agency for Research on Cancer classifies it as a 'possible' human carcinogen, and the EPA classifies DDT, DDE, and DDD as class B2 'probable' carcinogens. These evaluations are based mainly on the results of animal studies.

There is evidence from epidemiological studies (i.e. studies in human populations) that indicates that DDT causes cancers of the liver, pancreas and breast. There is mixed evidence that it contributes to leukemia, lymphoma and testicular cancer. Other epidemiological studies suggest that DDT/DDE does not cause multiple myeloma, or cancers of the prostate, endometrium, rectum, lung, bladder, or stomach.

### Breast Cancer

The question of whether DDT or DDE are risk factors of breast cancer has been repeatedly studied. While individual studies conflict, the most recent reviews of all the evidence conclude that pre-puberty exposure increases the risk of subsequent breast cancer. Until recently, almost all studies measured DDT or DDE blood levels at the time of breast cancer diagnosis or after. This study design has been criticized, since the levels at diagnosis do not necessarily correspond to levels when her cancer started. Taken as a whole such studies "do not support the hypothesis that exposure to DDT is an important risk factor for breast cancer." The studies of this design have been extensively reviewed.

In contrast, a study published in 2007 strongly associated early exposure (the *p,p'*- isomer) and breast cancer later in life. Unlike previous studies, this prospective cohort study collected blood samples from young mothers in the 1960s while DDT was still in use, and their breast cancer status was then monitored over the years. In addition to suggesting that the *p,p'*- isomer is the more significant risk factor, the study also suggests that the timing of exposure is critical. For the subset of women born more than 14 years before agricultural use, there was no association between DDT and breast cancer. However, for younger women—exposed earlier in life—the third who were exposed most to *p,p'*-DDT had a five-fold increase in breast cancer incidence over the least exposed third, after correcting for the protective effect of *o,p'*-DDT. These results are supported by animal studies.

### Use Against Malaria

Malaria remains a major public health challenge in many countries. 2008 WHO estimates were 243 million cases, and 863,000 deaths. About

89 per cent of these deaths occur in Africa, and mostly to children under the age of 5. DDT is one of many tools that public health officials use to fight the disease. Its use in this context has been called everything from a "miracle weapon [that is] like Kryptonite to the mosquitoes," to 'toxic colonialism'.

Before DDT, eliminating mosquito breeding grounds by drainage or poisoning with Paris green or pyrethrum was sometimes successful in fighting malaria. In parts of the world with rising living standards, the elimination of malaria was often a collateral benefit of the introduction of window screens and improved sanitation. Today, a variety of usually simultaneous interventions is the norm. These include antimalarial drugs to prevent or treat infection; improvements in public health infrastructure to quickly diagnose, sequester, and treat infected individuals; bednets and other methods intended to keep mosquitoes from biting humans; and vector control strategies such as larvaciding with insecticides, ecological controls such as draining mosquito breeding grounds or introducing fish to eat larvae, and indoor residual spraying with insecticides, possibly including DDT. IRS involves the treatment of all interior walls and ceilings with insecticides, and is particularly effective against mosquitoes, since many species rest on an indoor wall before or after feeding. DDT is one of 12 WHO-approved IRS insecticides. How much of a role DDT should play in this mix of strategies is still controversial.

WHO's anti-malaria campaign of the 1950s and 1960s relied heavily on DDT and the results were promising, though temporary. Experts tie the resurgence of malaria to multiple factors, including poor leadership, management and funding of malaria control programmes; poverty; civil unrest; and increased irrigation. The evolution of resistance to first-generation drugs (e.g. chloroquine) and to insecticides exacerbated the situation. Resistance was largely fueled by often unrestricted agricultural use. Resistance and the harm both to humans and the environment led many governments to restrict or curtail the use of DDT in vector control as well as agriculture.

Once the mainstay of anti-malaria campaigns, as of 2008 only 12 countries used DDT, including India and some southern African states, though the number is expected to rise.

## Effectiveness of DDT Against Malaria

When it was first introduced in World War II, DDT was very effective in reducing malaria morbidity and mortality. The WHO's anti-malaria campaign, which consisted mostly of spraying DDT, was initially very successful as well. For example, in Sri Lanka, the programme reduced cases from about 3 million per year before spraying to just 29 in 1964. Thereafter the programme was halted to save money and malaria rebounded to 600,000 cases in 1968 and the first quarter of 1969. The country resumed DDT vector control but the mosquitoes had acquired resistance in the interim, presumably because of continued agricultural use. The programme switched to malathion, which though more expensive proved effective.

Today, DDT remains on the WHO's list of insecticides recommended for IRS. Since the appointment of Arata Kochi as head of its anti-malaria division, WHO's policy has shifted from recommending IRS only in areas of seasonal or episodic transmission of malaria, to also advocating it in areas of continuous, intense transmission. The WHO has reaffirmed its commitment to eventually phasing DDT out, aiming "to achieve a 30 per cent cut in the application of DDT world-wide by 2014 and its total phase-out by the early 2020s if not sooner" while simultaneously combating malaria. The WHO plans to implement alternatives to DDT to achieve this goal.

South Africa is one country that continues to use DDT under WHO guidelines. In 1996, the country switched to alternative insecticides and malaria incidence increased dramatically. Returning to DDT and introducing new drugs brought malaria back under control. According to DDT advocate Donald Roberts, malaria cases increased in South America after countries in that continent stopped using DDT. Research data shows a significantly strong negative relationship between DDT residual house sprayings and malaria rates. In a research from 1993 to 1995, Ecuador increased its use of DDT and resulted in a 61 per cent reduction in malaria rates, while each of the other countries that gradually decreased its DDT use had large increase in malaria rates.

**Mosquito Resistance**

Resistance has greatly reduced DDT's effectiveness. WHO guidelines require that absence of resistance must be confirmed before using the chemical. Resistance is largely due to agricultural use, in much greater quantities than required for disease prevention. According to one study that attempted to quantify the lives saved by banning agricultural use and thereby slowing the spread of resistance, "it can be estimated that at current rates each kilo of insecticide added to the environment will generate 105 new cases of malaria".

Resistance was noted early in spray campaigns. Paul Russell, a former head of the Allied Anti-Malaria campaign, observed in 1956 that "resistance has appeared [after] six or seven years." DDT has lost much of its effectiveness in Sri Lanka, Pakistan, Turkey and Central America, and it has largely been replaced by organophosphate or carbamate insecticides, *e.g.* malathion or bendiocarb.

In many parts of India, DDT has also largely lost its effectiveness. Agricultural uses were banned in 1989, and its anti-malarial use has been declining. Urban use has halted completely. Nevertheless, DDT is still manufactured and used, and one study had concluded that "DDT is still a viable insecticide in indoor residual spraying owing to its effectivity in well supervised spray operation and high excito-repellency factor".

Studies of malaria-vector mosquitoes in KwaZulu-Natal Province, South Africa found susceptibility to 4 per cent DDT (the WHO susceptibility standard), in 63 per cent of the samples, compared to the average of 86.5 per cent in the same species caught in the open. The authors concluded that "Finding DDT resistance in the vector *An. arabiensis*, close to the area where we previously reported pyrethroid-resistance in the vector *An. funestus* Giles, indicates an urgent need to develop a strategy of insecticide resistance management for the malaria control programmes of southern Africa".

DDT can still be effective against resistant mosquitoes, and the avoidance of DDT-sprayed walls by mosquitoes is an additional benefit of the chemical. For example, a 2007 study reported that resistant mosquitoes avoided treated huts. The researchers argued that DDT was the best pesticide for use in IRS (even though it did not afford the most protection from mosquitoes out of the three test chemicals) because the other pesticides worked primarily by killing or irritating mosquitoes—encouraging the development of resistance to these agents. Others argue that the avoidance behaviour slows the eradication of the disease. Unlike other insecticides such as pyrethroids, DDT requires long exposure to accumulate a lethal dose; however its irritant property shortens contact periods. "For these reasons, when comparisons have been made, better malaria control has generally been achieved with pyrethroids than with DDT." In India, with its outdoor sleeping habits and frequent night duties, "the excito-repellent effect of DDT, often reported useful in other countries, actually promotes outdoor transmission.

### Residents' Concerns

For IRS to be effective, at least 80 per cent of homes and barns in an area must be sprayed. Lower coverage rates can jeopardize programme effectiveness. Many residents resist DDT spraying, objecting to the lingering smell, stains on walls, and may exacerbate problems with other insect pests. Pyrethroid insecticides (e.g. deltamethrin and lambda-cyhalothrin) can overcome some of these issues, increasing participation.

### Human Exposure

People living in areas where DDT is used for IRS have high levels of the chemical and its breakdown products in their bodies. Compared to contemporaries living where DDT is not used, South Africans living in sprayed homes have levels that are several orders of magnitude greater. Breast milk in regions where DDT is used against malaria greatly exceeds the allowable standards for breast-feeding infants. These levels are associated with neurological abnormalities in babies.

Most studies of DDT's human health effects have been conducted in developed countries where DDT is not used and exposure is relatively low. Many experts urge that alternatives be used instead of IRS. Epidemiologist Brenda Eskenazi argues, "We know DDT can save lives by repelling and killing disease-spreading mosquitoes. But evidence suggests that people living in areas where DDT is used are exposed to very high levels of the pesticide. The only published studies on health effects conducted in these populations have shown profound effects on male fertility. Clearly, more research is needed on the health of populations where indoor residual spraying is occurring, but in the meantime, DDT should really be the last resort against malaria rather than the first line of defense".

Illegal diversion to agriculture is also a concern, as it is almost impossible to prevent, and its subsequent use on crops is uncontrolled. For example, DDT use is widespread in Indian agriculture, particularly mango production, and is reportedly used by librarians to protect books. Other example include Ethiopia, where DDT intended for malaria control is reportedly being used in coffee production, and Ghana where it is used for fishing." The residues in crops at levels unacceptable for export have been an important factor in recent bans in several tropical countries. Adding to this problem is a lack of skilled personnel and supervision.

## Criticism of Restrictions on DDT Use

Critics claim that restricting DDT in vector control have caused unnecessary deaths due to malaria. Estimates range from hundreds of thousands, to much higher figures. Robert Gwadz of the National Institutes of Health said in 2007, "The ban on DDT may have killed 20 million children." These arguments have been dismissed as 'outrageous' by former WHO scientist Socrates Litsios. May Berenbaum, University of Illinois entomologist, says, "to blame environmentalists who oppose DDT for more deaths than Hitler is worse than irresponsible." Investigative journalist Adam Sarvana and others characterize this notion as a 'myth' promoted principally by Roger Bate of the pro-DDT advocacy group Africa Fighting Malaria (AFM).

Criticisms of a DDT 'ban' often specifically reference the 1972 US ban (with the erroneous implication that this constituted a worldwide ban and prohibited use of DDT in vector control). Reference is often made to Rachel Carson's *Silent Spring* even though she never pushed for a ban on DDT.

John Quiggin and Tim Lambert wrote, "the most striking feature of the claim against Carson is the ease with which it can be refuted." Carson actually devoted a page of her book to considering the relationship between DDT and malaria, warning of the evolution of DDT resistance in mosquitoes and concluding:

> It is more sensible in some cases to take a small amount of damage in preference to having none for a time but paying for it in the long run by losing the very means of fighting [is the advice given in Holland by Dr Briejer in his capacity as director of the Plant Protection Service]. Practical advice should be "Spray as little as you possibly can" rather than "Spray to the limit of your capacity."

According to Amir Attaran and Roger Bate, many environmental groups fought against the public health exception in the 2001 Stockholm Convention, over the objections of third world governments and many malaria researchers. Attaran strongly objected to an outright ban, writing, "Environmentalists in rich, developed countries gain nothing from DDT, and thus small risks felt at home loom larger than health benefits for the poor tropics. More than 200 environmental groups, including Greenpeace, Physicians for Social Responsibility and the World Wildlife Fund, actively condemn DDT. . ."

It has also been alleged that donor governments and agencies have refused to fund DDT spraying, or made aid contingent upon not using DDT. According to a report in the *British Medical Journal*, use of DDT in Mozambique "was stopped several decades ago, because 80 per cent of the country's health budget came from donor funds, and donors refused to allow the use of DDT." Roger Bate asserts, "many countries have been coming under pressure from international health and environment agencies to give up DDT or face losing aid grants: Belize and Bolivia are on record admitting they gave in to pressure on this issue from".

The United States Agency for International Development (USAID) has been the focus of much criticism. While the agency is currently funding the use of DDT in some African countries, in the past it did not. When John Stossel accused USAID of not funding DDT because it wasn't 'politically correct', Anne Peterson, the agency's assistant administrator for global health, replied that "I believe that the strategies we are using are as effective as spraying with DDT . . . So, politically correct or not, I am very confident that what we are doing is the right strategy." USAID's Kent R. Hill states that the agency has been misrepresented: "USAID strongly supports spraying as a preventative measure for malaria and will support the use of DDT when it is scientifically sound and warranted." The Agency's website states that "USAID has never had a 'policy' as such either 'for' or 'against' DDT for IRS. The real change in the past two years has been a new interest and emphasis on the use of IRS in general–with DDT or any other insecticide–as an effective malaria prevention strategy in tropical Africa." The website further explains that in many cases alternative malaria control measures were judged to be more cost-effective that DDT spraying, and so were funded instead.

Advocates of increased use of DDT in IRS claim that alternative insecticides are more expensive, more toxic, or not as effective. As discussed above, susceptibility of mosquitoes to DDT varies geographically. The same is true for alternative insecticides, so its relative effectiveness varies. Toxicity and cost-effectiveness comparisons lack data. Relative insecticide costs vary by location and ease of access, the habits of the local mosquitoes, the degrees of resistance exhibited by the mosquitoes, and the habits and compliance of the population, among other factors. The choice of insecticide has little impact on the total cost of a round of spraying, since product costs are only a fraction of campaign costs. IRS coverage needs to be maintained throughout the malaria season, making DDT's relatively long life an important cost savings.

Organophosphate and carbamate insecticides, *e.g.* malathion and bendiocarb, respectively, are more expensive than DDT per kilogram and are applied at roughly the same dosage. Pyrethroids such as deltamethrin are also more expensive than DDT, but are applied more sparingly (0.02-0.3 g/m$^2$ vs 1-2 g/m$^2$), so the net cost per house is about the same over 6 months.

**Non-chemical Vector Control**

Before DDT, malaria was successfully eradicated or curtailed in several tropical areas by removing or poisoning mosquito breeding grounds and larva habitats, for example by filling or applying oil to standing water. These methods have seen little application in Africa for more than half a century.

The relative effectiveness of IRS (with DDT or alternative insecticides) versus other malaria control techniques (e.g. bednets or prompt access to anti-malarial drugs) varies greatly and is highly dependent on local conditions.

A WHO study released in January 2008 found that mass distribution of insecticide-treated mosquito nets and artemisinin–based drugs cut malaria deaths in half in Rwanda and Ethiopia, countries with high malaria burdens. IRS with DDT did not play an important role in mortality reduction in these countries.

Vietnam has enjoyed declining malaria cases and a 97 per cent mortality reduction after switching in 1991 from a poorly funded DDT-based campaign to a programme based on prompt treatment, bednets, and pyrethroid group insecticides.

In Mexico, effective and affordable chemical and non-chemical strategies against malaria have been so successful that the Mexican DDT manufacturing plant ceased production due to lack of demand.

While the increased numbers of malaria victims since DDT usage collapsed document its value, many other factors contributed to the rise in cases.

A review of fourteen studies on the subject in sub-Saharan Africa, covering insecticide-treated nets, residual spraying, chemoprophylaxis for children, chemoprophylaxis or intermittent treatment for pregnant women, a hypothetical vaccine, and changing front–line drug treatment, found decision making limited by the gross lack of information on the costs and effects of many interventions, the very small number of cost-effectiveness analyses available, the lack of evidence on the costs and effects of packages of measures, and the problems in generalizing or comparing studies that relate to specific settings and use different methodologies and outcome measures. The two cost-effectiveness estimates of DDT residual spraying examined were not found to provide an accurate estimate of the cost-effectiveness of DDT spraying; furthermore, the resulting estimates may not be good predictors of cost-effectiveness in current programmes.

However, a study in Thailand found the cost per malaria case prevented of DDT spraying ($1.87 US) to be 21 per cent greater than the cost per case prevented of lambda-cyhalothrin–treated nets ($1.54 US), at very least casting some doubt on the unexamined assumption that DDT was the most cost-effective measure to use in all cases. The director of Mexico's malaria control programme finds similar results, declaring that it is 25 per cent cheaper for Mexico to spray a house with synthetic pyrethroids than with DDT. However, another study in South Africa found generally lower costs for DDT spraying than for impregnated nets.

A more comprehensive approach to measuring cost-effectiveness or efficacy of malarial control would not only measure the cost in dollars of the project, as well as the number of people saved, but would also consider ecological damage and negative aspects of insecticide use on human health. One preliminary study regarding the effect of DDT found that it is likely the detriment to human health approaches or exceeds the beneficial reductions in malarial cases, except perhaps in malarial epidemic situations. It is similar to the earlier mentioned study regarding estimated theoretical infant mortality caused by DDT and subject to the criticism also mentioned earlier.

A study in the Solomon Islands found that "although impregnated bed nets cannot entirely replace DDT spraying without substantial increase in incidence, their use permits reduced DDT spraying".

A comparison of four successful programmes against malaria in Brazil, India, Eritrea, and Vietnam does not endorse any single strategy but instead states, "Common success factors included conducive country conditions, a targeted technical approach using a package of effective tools, data-driven decision-making, active leadership at all levels of government, involvement of communities, decentralized implementation and control of finances, skilled technical and managerial capacity at national and sub-national levels, hands-on technical and programmemate support from partner agencies, and sufficient and flexible financing".

# Chapter 4
# Biodegradation

It is the chemical breakdown of materials by environment. The term is often used in relation to ecology, waste management and natural environmenta] (bioremediation). Organic material can be degraded aerobically with oxygen, or anaerobically, without oxygen. A term related to biodegradation is biomineralisation, in which organic matter is converted into minerals. Biosurfactant, an extracellular surfactant secreted by microorganisms, enhances the biodegradation process.

Biodegradable matter is generally organic material such as plant and animal matter and other substances originating from living organisms, or artificial materials that are similar enough to plant and animal matter to be put to use by microorganisms. Some microorganisms have a naturally occurring, microbial catabolic diversity to degrade, transform or accumulate a huge range of compounds including hydrocarbons (e.g. oil), polychlorinated biphenyls (PCBs), polyaromatic hydrocarbons (PAHs), pharmaceutical substances, radionuclides and metals. Major methodological breakthroughs in microbial biodegradation have enabled detailed genomic, metagenomic, proteomic, bioinformatic and other high-throughput analyses of environmentally relevant microorganisms providing unprecedented insights into key biodegradative pathways and the ability of microorganisms to adapt to changing environmental conditions.

Products that contain biodegradable matter and non-biodegradable matter are often marketed as biodegradable.

Biodegradation can be measured in a number of ways. The activity of aerobic microbes can be measured by the amount of oxygen they consume or the amount of carbon dioxide they produce. It can be measured by anaerobic microbes and the amount of methane or alloy that they may be able to produce. In formal scientific literature, the process is termed bio-remediation.

## Plastics

Biodegradable plastics. There are two main types of biodegradable plastics in the market: hydro-biodegradable plastics (HBP) and oxo-biodegradable plastics (OBP). Both will first undergo chemical degradation by hydrolysis and oxidation respectively. This results in their physical disintegration and a drastic reduction in their molecular weight. These smaller, lower molecular weight fragments are then amenable to biodegradation.

OBPs are made by adding a small proportion of compounds of specific transition metals (iron, manganese, cobalt and nickel are commonly used) into the normal production of polyolefins such as polyethylene (PE), polypropylene (PP) and polystyrene (PS). The additives act as catalysts to speed up the normal oxidative degradation, increasing the overall process by up to several orders of magnitude (factors of 10).

The products of the catalyzed oxidative degradation of the polyolefins are precisely the same as for conventional polyolefins because, other than a small amount of additive present, the plastics are conventional polyolefins. Many commercially useful hydrocarbons (e.g., cooking oils, polyolefins, many other plastics) contain small amounts of additives called antioxidants that prevent oxidative degradation during storage and use. Antioxidants function by 'deactivating' the free radicals that cause degradation. Lifetime (shelf life + use life) is controlled by antioxidant level and the rate of degradation after disposal is controlled by the amount and nature of the catalyst.

Since there are no existing corresponding standards that can be used directly in reference to plastics that enter the environment in other ways other than compost — i.e. as terrestrial or marine litter or in landfills, OBP technology is often attacked by the HBP industry as unable to live up to the standards (which are actually the standards for composting). It has to be understood that composting and biodegradation are not identical. OBP can however be tested according to ASTM D6954.

HBPs tend to degrade and biodegrade somewhat more quickly than OBP, but they have to be collected and put into an industrial composting unit. The end result is the same - both are converted to carbon dioxide ($CO_2$), water ($H_2O$) and biomass. OBP are generally less expensive, possess better physical properties and can be made with current plastics processing equipment. HBP emits methane in anaerobic conditions, but OBP does not.

Polyesters play a predominant role in hydro-biodegradable plastics due to their potentially dydrolysable ester bonds. HBP can be made from agricultural resources such as corn, wheat, sugar cane, or fossil (petroleum-based)resources , or blend of the two. Some of the commonly-used polymers include PHA (polyhydroxyalkanoates), PHBV (polyhydroxybutyrate-valerate), PLA (polylactic acid), PCL (polycaprolactone), PVA (polyvinyl alcohol), PET (polyethylene terephthalate) etc. It would be misleading to call these 'renewable' because the agricultural production process burns significant amounts of hydrocarbons and emits significant amounts of $CO_2$. OBPs (like normal plastics)are made from a by-product of oil or natural gas, which would be produced whether or not the by-product were used to make plastic.

HBP technology claims to be biodegradable by meeting the ASTM D6400-04 and EN 13432 Standards. However, these two commonly quoted standards are related to the performance of plastics in a commercially managed compost environment. They are not biodegradation standards. Both were developed for hydro-biodegradable polymers where the mechanism including biodegradation is based on reaction with water and state that in order for a production to be compostable, the following criteria need to be met:

- *Disintegration*, the ability to fragment into non-distinguishable pieces after screening and safely support bio-assimilation and microbial growth;
- *Inherent biodegradation*, conversion of carbon to carbon dioxide to the level of 60 per cent and 90 per cent over a period of 180 days for ASTM D6400-04 and EN 13432 respectively; There is therefore little or no carbon left for the benefit of the soil, but the $CO_2$ emitted to atmosphere contributes to climate-change;
- *Safety*, that there is no evidence of any eco-toxicity in finished compost and soils can support plant growth; and
- *Toxicity*, that heavy metal concentrations are less than 50 per cent regulated values in soil amendments.

## Biodegradable Plastic

Biodegradable plastics are plastics that will decompose in natural aerobic (composting) and anaerobic (landfill) environments. Biodegradation of plastics can be achieved by enabling microorganisms in the environment to metabolize the molecular structure of plastic films to produce an inert humus-like material that is less harmful to the environment. They may be composed of either bioplastics, which are plastics whose components are derived from renewable raw materials, or petroleum-based plastics which utilize an additive. The use of bio-active compounds compounded with swelling agents ensures that, when combined with heat and moisture, they expand the plastic's molecular structure and allow the bio-active compounds to metabolize and neutralize the plastic.

Biodegradable plastics typically are produced in two forms: injection molded (solid, 3D shapes), typically in the form of disposable food service items, and films, typically organic fruit packaging and collection bags for leaves and grass trimmings, and agricultural mulch.

In the United States, ASTM International is the authoritative body for defining biodegradable standards. The specific subcommittee responsibility for overseeing these standards falls on the Committee D20.96 on Environmentally Degradable Plastics and Biobased Products. The current ASTM standards are defined as standard specifications and standard test methods. Standard specifications create a pass or fail scenario whereas standard test methods identify the specific testing parameters for facilitating specific biodegradable tests on plastics.

Currently, there are three such ASTM standard specifications which mostly address biodegradable plastics in composting type environments, the ASTM D6400-04 Standard Specification for Compostable Plastics, ASTM D6868 - 03 Standard Specification for Biodegradable Plastics Used as Coatings on Paper and Other Compostable Substrates, and the ASTM D7081 — 05 Standard Specification for Non-Floating Biodegradable Plastics in the Marine Environment.

Currently the most accurate standard test method for anaerobic environments is the ASTM D5511-02 Standard Test Method for Determining Anaerobic Biodegradation of Plastic Materials Under High-Solids Anaerobic-Digestion Conditions. Another standard test method for testing in anaerobic environments is the ASTM D5526-94(2002) Standard Test Method for Determining Anaerobic Biodegradation of Plastic Materials Under Accelerated Landfill Conditions, this test has proven extremely difficult to perform.

The current California legislation AB 1972 ensures accurate environmental advertising of plastics by allowing only the use of terms that can be verified by an American Society for Testing Materials (ASTM) standard specification. This legislation does not include ASTM standard test methods. The two ASTM standard specifications which are used in the legislation are ASTM D6400 and D7081. Products passing these ASTM specifications can use the term compostable on the product label.

Biodegradable plastics are not a panacea, however. Some critics claim that a potential environmental disadvantage of certified biodegradable plastics is that the carbon that is locked up in them is released into the atmosphere as a greenhouse gas. However, biodegradable plastics from natural materials, such as vegetable crop derivatives or animal products, sequester $CO_2$ during the phase when they're growing, only to release $CO_2$ when they're decomposing, so there is no net gain in carbon dioxide emissions.

However, certified biodegradable plastics require a specific environment of moisture and oxygen to biodegrade, conditions found in professionally managed composting facilities. There is much debate about the total carbon, fossil fuel and water usage in processing biodegradable plastics from natural materials and whether they are a negative impact to human food supply. Traditional plastics made from non-renewable fossil fuels lock up much of the carbon in the plastic as opposed to being utilized in the processing of the plastic. The carbon is permanently trapped inside the plastic lattice, and is rarely recycled.

There is concern that another greenhouse gas, methane, might be released when any biodegradable material, including truly biodegradable plastics, degrades in an anaerobic (landfill) environment. Methane production from landfills is rarely captured or burned, but rather enter the atmosphere, where it is a potent greenhouse gas. Methane production from specially managed landfill environments is captured and burned to reduce the release of methane in the environment. Some landfills today capture the methane biogas for use in clean inexpensive energy. Of course, incinerating non-biodegradable plastics will release carbon dioxide as well. Disposing of biodegradable plastics made from natural materials in anaerobic (landfill) environments will result in the plastic lasting for hundred of years.

The US EPA has mandated strict standards for landfill design and construction to prevent biodegradation in a landfill in the first place. The intentional production of methane from landfills is, therefore, the rare exception and not the rule for most municipal solid waste.

It is also possible that bacteria will eventually develop the ability to degrade plastics. This has already happened with nylon: two types of nylon eating bacteria, *Flavobacteria* and *Pseudomonas*, were found in 1975 to possess enzymes (nylonase) capable of breaking down nylon. While not a solution to the disposal problem, it is likely that bacteria will evolve the ability to use other synthetic plastics as well. In 2008, a 16-year-old boy reportedly isolated two plastic-consuming bacteria.

The latter possibility was in fact the subject of a cautionary novel by Kit Pedler and Gerry Davis (screenwriter), the creators of the Cybermen, re-using the plot of the first episode of their Doomwatch series. The novel,

*Mutant 59: The Plastic Eater*, written in 1971, is the story of what could happen if a bacterium were to evolve–or be artificially cultured–to eat plastics, and be let loose in a major city.

Materials such as a polyhydroxyalkanoate (PHA) biopolymer are completely compostable in an industrial compost facility. Polylactic acid (PLA) is another 100 per cent compostable biopolymer which can fully degrade above 60C in an industrial composting facility. Fully biodegradable plastics are more expensive, partly because they are not widely enough produced to achieve large economies of scale.

Certain additives when added to conventional plastics attract the microbes to the molecular structure by allowing the hydrocarbons to be sensed once again by microbial colonies. When oil is in the ground, the microbes attach themselves onto the hydrocarbons consuming the oil and creating natural gas, 50 per cent of which is methane gas. When the oil is cracked 4 per cent is used for the plastic industry, if the plastic industry did not use this 4 per cent the 4 per cent would be considered waste and be thrown away or removed and dumped into a waste disposal facility, another 4 per cent is used in the generation of your consumer product. During this phase of cracking the organic compound which attracts the microbes to the molecular structure of the plastic is burnt out. The organic compound which is burnt out and other proprietary compounds which increase quorum sensing of the microbes and pH balance for the microbes are placed into the molecular structure of the plastic, to create a plastic product that can biodegrade 100 times faster than normal plastic.

Over 200 million tons of plastic are manufactured annually around the world, according to the Society of Plastics Engineers. Of those 200 million tons, 26 million are manufactured in the United States. The EPA reported in 2003 that only 5.8 per cent of those 26 million tons of plastic waste are recycled, although this is increasing rapidly.

Much of the reason for disappointing plastics recycling goals is that conventional plastics are often commingled with organic wastes (food scraps, wet paper, and liquids), making it difficult and impractical to recycle the underlying polymer without expensive cleaning and sanitizing procedures.

On the other hand, composting of these mixed organics (food scraps, yard trimmings, and wet, non-recyclable paper) is a potential strategy for recovering large quantities of waste and dramatically increase community recycling goals. Food scraps and wet, non-recyclable paper comprises 50 million tons of municipal solid waste. Biodegradable plastics can replace the non-degradable plastics in these waste streams, making municipal composting a significant tool to divert large amounts of otherwise nonrecoverable waste from landfills.

If even a small amount of conventional plastics were to be commingling with organic materials, the entire batch of organic waste is 'contaminated' with small bits of plastic that spoil prime-quality compost humus. Composters, therefore, will not accept mixed organic waste streams unless they are completely devoid of nondegradable plastics. So, because of a relatively small quantity of non-degradable plastics, a significant waste disposal strategy is stalled.

However, proponents of biodegradable plastics argue that these materials offer a solution to this problem. Certified biodegradable plastics combine the utility of plastics (lightweight, resistance, relative low cost) with the ability to completely and fully biodegrade in a compost facility. Rather than worrying about recycling a relatively small quantity of commingled plastics, these proponents argue that certified biodegradable plastics can be readily commingled with other organic wastes, thereby enabling composting of a much larger position of nonrecoverable solid waste. Commercial composting for all mixed organics then becomes commercially viable and economically sustainable. More municipalities can divert significant quantities of waste from overburdened landfills since the entire waste stream is now biodegradable and therefore easier to process.

The use of biodegradable plastics, therefore, is seen as an enabler for the complete recovery of large quantities of municipal sold waste (via aerobic composting) that were are heretofore unrecoverable by other means except land filling or incineration.

## Confusion Over Proper Definition of Terms

Until recently there were few legal standards regarding marketing claims surrounding the use of the term 'biodegradable'. In 2007, the state of California passed regulation banning companies from claiming their products are biodegradable without proper scientific certification from a third-party laboratory.

The Federal Court of Australia declared on March 30, 2009 that a director of a company that manufactured 'biodegradable' disposable diapers (who also approved the company's advertising) had been knowingly making false and misleading claims about biodegradability.

In June 2009, the Federal Trade Commission charged two companies with making unsupported marketing claims regarding biodegradability.

## Energy Costs for Production

Various researchers have undertaken extensive life cycle assessments of biodegradable polymers to determine whether these materials are more energy efficient than polymers made by conventional fossil fuel-based means. Research done by Gerngross, *et al.* estimates that the fossil fuel energy required to produce a kilogram of polyhydroxyalkanoate (PHA) is 50.4 MJ/kg, which coincides with another estimate by Akiyama, *et al.* , who estimate a value between 50-59 MJ/kg. This information does not take into account the feedstock energy, which can be obtained from non-fossil fuel based methods. Polylactide (PLA) was estimated to have a fossil fuel energy cost of 54-56.7 from two sources , but recent developments in the commercial production of PLA by NatureWorks has eliminated some dependence fossil fuel based energy by supplanting it with wind power and biomass-driven strategies. They report making a kilogram of PLA with only 27.2 MJ of fossil fuel-based energy and anticipate that this number will drop to 16.6 MJ/kg in their next generation plants. In contrast, polypropylene and high density polyethylene require 85.9 and 73.7 MJ/kg respectively, but these values include the embedded energy of the feedstock because it is based on fossil fuel.

Gerngross reports a 2.65 total fossil fuel energy equivalent (FFE) required to produce a single kilogram of PHA, while polypropylene only requires 2.2 kg FFE . Gerngross assesses that the decision to proceed forward with any biodegradable polymer alternative will need to take into account the priorities of society with regard to energy, environment, and economic cost.

Furthermore, it is important to realize the youth of alternative technologies. Technology to produce PHA, for instance, is still in development today, and energy consumption can be further reduced by eliminating the fermentation step, or by utilizing food waste as feedstock. The use of alternative crops other than corn, such as sugar cane from Brazil, are expected to lower energy requirements- manufacturing of PHAs by fermentation in Brazil enjoys a favourable energy consumption scheme where bagasse is used as source of renewable energy.

Many biodegradable polymers that come from renewable resources (i.e., starch-based, PHA, PLA) also compete with food production, as the primary feedstock is currently corn. For the US to meet its current output of plastics production with BPs, it would require 1.62 square metres per kilogram produced . While this space requirement could be feasible, it is always important to consider how much impact this large scale production could have on food prices and the opportunity cost of using land in this fashion versus alternatives.

Before Flue Gas Desulfurization was Installed, the Emissions from This Power Plant in New Mexico Contained Excessive Amounts of Sulfur Dioxide

Raw Sewage and Industrial Waste Flows Across International Borders—New River Passes from Mexicali to Calexico, California

Excavation Showing Soil Contamination at a Disused Gasworks

A Boeing 747-400 Passes Close to Houses Shortly Before Landing at London Heathrow Airport

The Objective of Sewage Treatment is to Produce a Disposable Effluent Without Causing Harm to the Surrounding Environ-ment and Prevent Pollution

Secondary Sedimentation Tank at a Rural Treatment Plant

Defective and Obsolete Electronic Equipments

A Landfill Compaction Vehicle in Action

# Sewage

Sewage is water-carried wastes, in either solution or suspension, that is intended to flow away from a community. Also known as wastewater flows, sewage is the used water supply of the community. It is more than 99.9 per cent pure water and is characterized by its volume or rate of flow, its physical condition, its chemical constituents, and the bacteriological organisms that it contains. Depending on their origin, wastewater can be classed as sanitary, commercial, industrial, agricultural or surface runoff.

The spent water from residences and institutions, carrying body wastes, washing water, food preparation wastes, laundry wastes, and other waste products of normal living, are classed as domestic or sanitary sewage. Liquid-carried wastes from stores and service establishments serving the immediate community, termed commercial wastes, are included in the sanitary or domestic sewage category if their characteristics are similar to household flows. Wastes that result from an industrial process or the production or manufacture of goods are classed as industrial wastes. Their flows and strengths are usually more varied, intense, and concentrated than those of sanitary sewage. Surface runoff, also known as storm flow or overland flow, is that portion of precipitation that runs rapidly over the ground surface to a defined channel. Precipitation absorbs gases and particulates from the atmosphere, dissolves and leaches materials from vegetation and soil, suspends matter from the land, washes spills and debris from urban streets and highways, and carries all these pollutants as wastes in its flow to a collection point.

Wastewater from all of these sources may carry pathogenic organisms that can transmit disease to humans and other animals; contain organic matter that can cause odor and nuisance problems; hold nutrients that may cause eutrophication of receiving water bodies; and can lead to ecotoxicity. Proper collection and safe, nuisance-free disposal of the liquid wastes of a community are legally recognized as a necessity in an urbanized, industrialized society.

'Sewage' and 'Sewerage' may be used interchangeably in the USA but elsewhere they retain separate and different meanings — sewage being the liquid material and sewerage being the pipes, pumps and infrastructure through which sewage flows.

A system of sewer pipes (sewers) collects sewage and takes it for treatment or disposal. The system of sewers is called *sewerage* or *sewerage system* in British English and *sewage system* in American English. Where a main sewerage system has not been provided, sewage may be collected from homes by pipes into septic tanks or cesspits, where it may be treated or collected in vehicles and taken for treatment or disposal. Properly functioning septic tanks require emptying every 2-5 years depending on the load of the system.

Sewage and waste water is also disposed of to rivers, streams and the sea in many parts of the world. Doing so can lead to serious pollution of the receiving water. This is common in third world countries and may still occur in some developed countries, where septic tank systems are too expensive.

## Treatment

Sewage treatment is the process of removing the contaminants from sewage to produce liquid and solid (sludge) suitable for discharge to the environment or for reuse. It is a form of waste management. A septic tank or other on-site wastewater treatment system such as biofilters can be used to treat sewage close to where it is created.

Sewage water is a complex matrix, with many distinctive chemical characteristics. These include high concentrations of ammonium, nitrate, phosphorus, high conductivity (due to high dissolved solids), high alkalinity, with pH typically ranging between 7 and 8. Trihalomethanes are also likely to be present as a result of past disinfection.

In developed countries sewage collection and treatment is typically subject to local, state and federal regulations and standards.

## Conversion to Fertiliser

Sewage sludge can be collected through a sludge processing plant that automatically heats the matter and conveys it into fertiliser pellets (hereby removing possible contamination by chemical detergents, . . .) This approach allows to eliminate seawater pollution by conveying the water directly to the sea without treatment (a practice which is still common in developing countries, despite environmental regulation). Sludge plants are useful in areas that have already set-up a sewage-system, but not in areas without such a system, as composting toilets are more efficient and do not require sewage pipes (which break over time).

## Sewage Collection and Disposal

Sewage collection and disposal systems transport sewage through cities and other inhabited areas to sewage treatment plants to protect public health and prevent disease. Sewage is treated to control water pollution before discharge to surface waters.

A sewage system may convey the wastewater by gravity to a sewage treatment plant. Where pipeline excavation is difficult because of rock or there is limited topographic relief (i.e., due to flat terrain), gravity collection systems may not be practical and the sewage must be pumped through a pipeline to the treatment plant. In low-lying communities, wastewater may be conveyed by vacuum. Pipelines range in size from pipes of six inches (150 mm) in diameter to concrete-lined tunnels of up to thirty feet (10 m) in diametre.

Community sewage can also be collected by an effluent sewer system, also known as a STEP system (Septic Tank Effluent Pumping). At each home, a buried collection tank is used to separate solids from the liquid effluent portion. Only the liquid portion is then pumped through small diameter pipe (typically 1.5" to 4") to downstream treatment. Because the wastestream is pressurized, the pipes can be laid just below the ground surface along the land's contour.

Sewage can also be collected by low pressure pumps and vacuum systems. A low pressure system uses a small grinder pump located at each point of connection, typically a house or business. Vacuum sewer systems use differential atmospheric pressure to move the liquid to a central vacuum station. Typically a vacuum sewer station can service approximately 1,200 homes before it becomes more cost-effective to build another station.

## Design and Analysis of Collection Systems

Design and sizing of sewage collection systems considers population served, commercial and industrial flows, flow peaking characteristics and wet weather flows. Combined sewer systems are designed to transport both stormwater runoff and sewage in the same pipe. Besides the projected sewage flow, the size and characteristics of the watershed are the overriding design considerations for combined sewers. Often, combined sewers can not handle the volume of runoff, resulting in combined sewer overflows and causing water pollution problems in nearby water bodies.

Separate sanitary sewer systems are designed to transport sewage alone. In communities served by separate sanitary sewers, another pipe system is constructed to convey stormwater runoff directly to surface waters. Most municipal sewer systems constructed today are separate sewer systems.

Although separate sewer systems are intended to transport only sewage, all sewer systems have some degree of inflow and infiltration of surface water and groundwater, which can lead to sanitary sewer overflows. Inflow and infiltration is highly affected by antecedent moisture conditions, which also represents an important design consideration in these systems.

A sewer bed is a piece of land typically used by a municipality for the dumping of raw sewage. Usually raw sewage was brought by truck or drawn by horses to be dumped, but the practice stopped back in the 1940s.

## Historical Sewage Conveyance and Disposal

The historical focus of sewage treatment was on conveyance of raw sewage to a natural body of water, e.g. a river or ocean, where it would be satisfactorily diluted and dissipated. Early human habitations were often built next to water sources. Rivers could double as a crude form of natural sewage disposal.

According to Teresi (2002):

> The Indus architects designed sewage disposal systems on a large scale, building networks of brick effluent drains following the lines of the streets. The drains were seven to ten feet wide, cut at two feet below ground level with U-shaped bottoms lined with loose brick easily taken up for cleaning. At the intersection of two drains, the sewage planners installed cesspools with steps leading down into them, for periodic cleaning. By 2700 B.C., these cities had standardized earthenware plumbing pipes with broad flanges for easy joining with asphalt to stop leaks.

## Ancient Systems

The first sanitation system has been found at the prehistoric Middle East, in south-east of Iran near Zabol in Burnt City (Shahre soukhteh) areas. The first time an inverted siphon system was used, along with glass covered clay pipes, was in the palaces of Crete, Greece. It is still in working condition, after about 3000 years.

Higher population densities required more complex sewer collection and conveyance systems to maintain (somewhat) sanitary conditions in crowded cities. The ancient cities of Harappa and Mohenjo-daro of the Indus Valley civilization constructed complex networks of brick-lined sewage drains from around 2600 BC and also had outdoor flush toilets connected to this network.

The urban areas of the Indus Valley civilization provided public and private baths, sewage was disposed through underground drains built with precisely laid bricks, and a sophisticated water management system with numerous reservoirs was established. In the drainage systems, drains from houses were connected to wider public drains.

Ancient Minoan civilization had stone sewers that were periodically flushed with clean water.

Roman towns and garrisons in the United Kingdom between 46 BC and 400 AD had complex sewer networks sometimes constructed out of hollowed out Elm logs which were shaped so that they butted together with the down-stream pipe providing a socket for the upstream pipe.

A significant development was the construction of a network of sewers to collect waste water, which began from the Indus Valley civilization. In some cities, including Rome, Istanbul (Constantinople) and Fustat, networked ancient sewer systems continue to function today as collection systems for those cities' modernized sewer systems. Instead of flowing to a river or the sea, the pipes have been re-routed to modern sewer treatment facilities.

## 16th Century

The system then remained with not much progress until the 16th century, where, in England, Sir John Harington invented a device for Queen Elizabeth (his godmother) that released wastes into cesspools.

However, many cities had no sewers and relied on nearby rivers or occasional rain to wash away sewage. In some cities, waste water simply ran down the streets, which had stepping stones to keep pedestrians out of the muck, and eventually drained as runoff into the local watershed. This was enough in early cities with few occupants but the growth of cities quickly overpolluted streets and became a constant source of disease. Even as recently as the late 19th century sewerage systems in parts of the highly industrialised United Kingdom were so inadequate that water-borne diseases such as cholera and typhoid were still common. In Merthyr Tydfil, a large town in South Wales, most houses discharged their sewage to individual cess-pits which persistently overflowed causing the pavements to be awash with foul sewage.

## Industrial Revolution Era

As an outgrowth of the Industrial Revolution, many cities in Europe and North America grew in the 19th century, frequently leading to crowding and increasing concerns about public health. As part of a trend of municipal sanitation programmes in the late 19th and 20th centuries, many cities constructed extensive sewer systems to help control outbreaks of disease. :29-34 Initially these systems discharged sewage directly to surface waters without treatment. The first comprehensive sewer system was built in Hamburg, Germany in the mid-19th century. The first such systems in the United States were built in the late 1850s in Chicago and Brooklyn.

As pollution of water bodies became a concern, cities attempted to treat the sewage before discharge. Early techniques involved land application of sewage on agricultural land. In the late 19th century some cities began to add chemical treatment and sedimentation systems to their sewers. In the United States, the first sewage treatment plant using chemical precipitation was built in Worcester, Massachusetts in 1890.

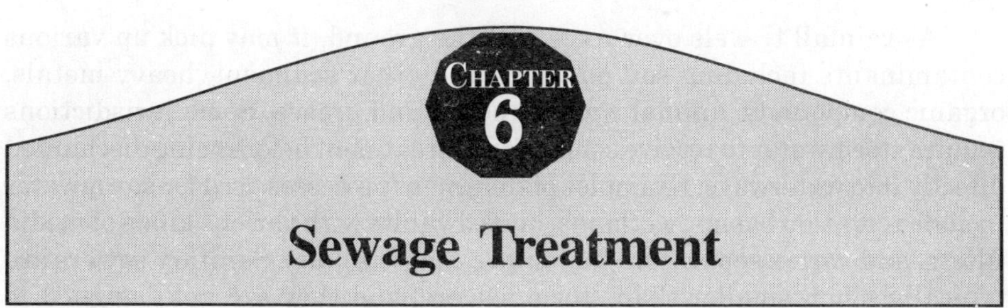

# Chapter 6

# Sewage Treatment

Sewage treatment, or domestic wastewater treatment, is the process of removing contaminants from wastewater and household sewage, both runoff (effluents) and domestic. It includes physical, chemical, and biological processes to remove physical, chemical and biological contaminants. Its objective is to produce an environmentally-safe fluid waste stream (or treated effluent) and a solid waste (or treated sludge) suitable for disposal or reuse (usually as farm fertilizer). Using advanced technology it is now possible to re-use sewage effluent for drinking water, although Singapore is the only country to implement such technology on a production scale in its production of NE Water.

Sewage is created by residential, institutional, and commercial and industrial establishments and includes household waste liquid from toilets, baths, showers, kitchens, sinks and so forth that is disposed of via sewers. In many areas, sewage also includes liquid waste from industry and commerce.

The separation and draining of household waste into greywater and blackwater is becoming more common in the developed world, with greywater being permitted to be used for watering plants or recycled for flushing toilets. Most sewage also includes some surface water from roofs or hard-standing areas and may include stormwater runoff.

Sewerage systems capable of handling stormwater are known as combined systems or combined sewers. Such systems are usually avoided now since they complicate and thereby reduce the efficiency of sewage treatment plants owing to their seasonality. The wide variability in flow, affected by precipitation, also leads to a need to construct much larger, more expensive, treatment facilities than would otherwise be required. In addition, heavy storms that contribute greater excess flow than the treatment plant can handle may overwhelm the sewage treatment system, causing a spill or overflow. Modern sewered developments tend to be provided with separate storm drain systems for rainwater.

As rainfall travels over roofs and the ground, it may pick up various contaminants including soil particles and other sediment, heavy metals, organic compounds, animal waste, and oil and grease. Some jurisdictions require stormwater to receive some level of treatment before being discharged directly into waterways. Examples of treatment processes used for stormwater include retention basins, wetlands, buried vaults with various kinds of media filters, and vortex separators (to remove coarse solids). Sanitary sewers are typically much smaller than storm sewers, and they are not designed to transport stormwater. In areas with basements, backups of raw sewage can occur if excessive stormwater is allowed into a sanitary sewer system.

Sewage can be treated close to where it is created, a decentralised system, (in septic tanks, biofilters or aerobic treatment systems), or be collected and transported via a network of pipes and pump stations to a municipal treatment plant, a centralised system, (see sewerage and pipes and infrastructure). Sewage collection and treatment is typically subject to local, state and federal regulations and standards. Industrial sources of wastewater often require specialized treatment processes.

Sewage treatment generally involves three stages, called primary, secondary and tertiary treatment.

1. *Primary treatment* consists of temporarily holding the sewage in a quiescent basin where heavy solids can settle to the bottom while oil, grease and lighter solids float to the surface. The settled and floating materials are removed and the remaining liquid may be discharged or subjected to secondary treatment.

2. *Secondary treatment* removes dissolved and suspended biological matter. Secondary treatment is typically performed by indigenous, water-borne micro-organisms in a managed habitat. Secondary treatment may require a separation process to remove the micro-organisms from the treated water prior to discharge or tertiary treatment.

3. *Tertiary treatment* is sometimes defined as anything more than primary and secondary treatment in order to allow rejection into a highly sensitive or fragile ecosystem (estuaries, low-flow rivers, coral reefs, . . .). Treated water is sometimes disinfected chemically or physically

(for example, by lagoons and microfiltration) prior to discharge into a stream, river, bay, lagoon or wetland, or it can be used for the irrigation of a golf course, green way or park. If it is sufficiently clean, it can also be used for groundwater recharge or agricultural purposes.

### Pre-treatment

Pre-treatment removes materials that can be easily collected from the raw wastewater before they damage or clog the pumps and skimmers of primary treatment clarifiers (trash, tree limbs, leaves, etc.).

### Screening

The influent sewage water is screened to remove all large objects carried in the sewage stream. This is most commonly done with an automated mechanically raked bar screen in modern plants serving large populations, whilst in smaller or less modern plants a manually cleaned screen may be used. The raking action of a mechanical bar screen is typically paced according to the accumulation on the bar screens and/or flow rate. The solids are collected and later disposed in a landfill or incinerated. Bar screens or mesh screens of varying sizes may be used to optimise solids removal. If gross solids are not removed they become entrained in pipes and moving parts of the treatment plant and can cause substantial damage and inefficiency in the process.

### Grit Removal

Pre-treatment may include a sand or grit channel or chamber where the velocity of the incoming wastewater is adjusted to allow the settlement of sand, grit, stones, and broken glass. These particles are removed because they may damage pumps and other equipment. For small sanitary sewer systems, the grit chambers may not be necessary, but grit removal is desirable at larger plants.

### Fat and Grease Removal

In some larger plants, fat and grease is removed by passing the sewage through a small tank where skimmers collect the fat floating on the surface. Air blowers in the base of the tank may also be used to help recover the fat as a froth. In most plants however, fat and grease removal takes place in the primary settlement tank using mechanical surface skimmers.

### Primary Treatment

In the primary sedimentation stage, sewage flows through large tanks, commonly called 'primary clarifiers' or "primary sedimentation tanks." The tanks are used to settle sludge while grease and oils rise to the surface and are skimmed off. Primary settling tanks are usually equipped with mechanically driven scrapers that continually drive the collected sludge towards a hopper in the base of the tank where it is pumped to sludge treatment facilities. Grease and oil from the floating material can sometimes be recovered for saponification.

The dimensions of the tank should be designed to effect removal of a high percentage of the floatables and sludge. A typical sedimentation tank may remove from 60 to 65 per cent of suspended solids, and from 30 to 35 per cent of biochemical oxygen demand (BOD) from the sewage.

## Secondary Treatment

Secondary treatment is designed to substantially degrade the biological content of the sewage which are derived from human waste, food waste, soaps and detergent. The majority of municipal plants treat the settled sewage liquor using aerobic biological processes. To be effective, the biota require both oxygen and food to live. The bacteria and protozoa consume biodegradable soluble organic contaminants (e.g. sugars, fats, organic short-chain carbon molecules, etc.) and bind much of the less soluble fractions into floc. Secondary treatment systems are classified as *fixed-film* or *suspended-growth* systems.

Fixed-film or attached growth systems include trickling filters and rotating biological contactors, where the biomass grows on media and the sewage passes over its surface.

Suspended-growth systems include activated sludge, where the biomass is mixed with the sewage and can be operated in a smaller space than fixed-film systems that treat the same amount of water. However, fixed-film systems are more able to cope with drastic changes in the amount of biological material and can provide higher removal rates for organic material and suspended solids than suspended growth systems.

Roughing filters are intended to treat particularly strong or variable organic loads, typically industrial, to allow them to then be treated by conventional secondary treatment processes. Characteristics include filters filled with media to which wastewater is applied. They are designed to allow high hydraulic loading and a high level of aeration. On larger installations, air is forced through the media using blowers. The resultant wastewater is usually within the normal range for conventional treatment processes.

A filter removes a small percentage of the suspended organic matter, while the majority of the organic matter undergoes a change of character, only due to the biological oxidation and nitrification taking place in the filter. With this aerobic oxidation and nitrification, the organic solids are converted into coagulated suspended mass, which is heavier and bulkier, and can settle to the bottom of a tank. The effluent of the filter is therefore passed through a sedimentation tank, called a secondary clarifier, secondary settling tank or humus tank.

In general, activated sludge plants encompass a variety of mechanisms and processes that use dissolved oxygen to promote the growth of biological floc that substantially removes organic material.

The process traps particulate material and can, under ideal conditions, convert ammonia to nitrite and nitrate and ultimately to nitrogen gas. Surface-aerated basins (Lagoons)

Many small municipal sewage systems in the United States (1 million gal./day or less) use aerated lagoons.

Most biological oxidation processes for treating industrial wastewaters have in common the use of oxygen (or air) and microbial action. Surface-aerated basins achieve 80 to 90 per cent removal of BOD with retention times of 1 to 10 days. The basins may range in depth from 1.5 to 5.0 metres and use motor-driven aerators floating on the surface of the wastewater.

In an aerated basin system, the aerators provide two functions: they transfer air into the basins required by the biological oxidation reactions, and they provide the mixing required for dispersing the air and for contacting the reactants (that is, oxygen, wastewater and microbes). Typically, the floating surface aerators are rated to deliver the amount of air equivalent to 1.8 to 2.7 kg $O_2$/kW·h. However, they do not provide as good mixing as is normally achieved in activated sludge systems and therefore aerated basins do not achieve the same performance level as activated sludge units.

Biological oxidation processes are sensitive to temperature and, between 0° C and 40° C, the rate of biological reactions increase with temperature. Most surface aerated vessels operate at between 4° C and 32° C.

## Constructed Wetlands

Constructed wetlands (can either be surface flow or subsurface flow, horizontal or vertical flow), include engineered reedbeds and belong to the family of phytorestoration and ecotechnologies; they provide a high degree of biological improvement and depending on design, act as a primary, secondary and sometimes tertiary treatment, also see phytoremediation. One example is a small reedbed used to clean the drainage from the elephants' enclosure at Chester Zoo in England; numerous CWs are used to recycle the water of the city of Honfleur in France and numerous other towns in Europe, the US, Asia and Australia. They are known to be highly productive systems as they copy natural wetlands, called the 'Kidneys of the earth' for their fundamental recycling capacity of the hydrological cycle in the biosphere. Robust and reliable, their treatment capacities improve as time go by, at the opposite of conventional treatment plants whose machinery age with time. They are being increasingly used, although adequate and experienced design are more fundamental than for other systems and space limitation may impede their use.

### Filter Beds (Oxidizing Beds)

In older plants and those receiving variable loadings, trickling filter beds are used where the settled sewage liquor is spread onto the surface of a bed made up of coke (carbonized coal), limestone chips or specially fabricated plastic media. Such media must have large surface areas to support the biofilms that form. The liquor is typically distributed through perforated spray arms. The distributed liquor trickles through the bed and is collected in drains at the base. These drains also provide a source of air which percolates up through the bed, keeping it aerobic. Biological films of bacteria, protozoa and fungi form on the media's surfaces and eat or otherwise reduce the organic content. This biofilm is often grazed by insect larvae, snails, and worms which help maintain an optimal thickness. Overloading of beds increases the thickness of the film leading to clogging of the filter media and ponding on the surface. Recent advances in media and process micro-biology design overcome many issues with Trickling filter designs.

### Soil Bio-Technology

A new process called Soil Bio-Technology (SBT) developed at IIT Bombay has shown tremendous improvements in process efficiency enabling total water reuse, due to extremely low operating power requirements of less than 50 joules per kg of treated water. Typically SBT systems can achieve chemical oxygen demand (COD) levels less than 10 mg/L from sewage input of COD 400 mg/L. SBT plants exhibit high reductions in COD values and bacterial counts as a result of the very high microbial densities available in the media. Unlike conventional treatment plants, SBT plants produce insignificant amounts of sludge, precluding the need for sludge disposal areas that are required by other technologies.

In the Indian context, conventional sewage treatment plants fall into systemic disrepair due to:

1. high operating costs;
2. equipment corrosion due to methanogenesis and hydrogen sulphide;
3. non-reusability of treated water due to high COD (>30 mg/L) and high fecal coliform (>3000 NFU) counts;
4. lack of skilled operating personnel; and
5. equipment replacement issues.

Examples of such systemic failures has been documented by Sankat Mochan Foundation at the Ganges basin after a massive clean-up effort by the Indian government in 1986 by setting up sewage treatment plants under the Ganga Action Plan failed to improve river water quality.

## Biological Aerated Filters

Biological Aerated (or Anoxic) Filter (BAF) or Biofilters combine filtration with biological carbon reduction, nitrification or denitrification. BAF usually includes a reactor filled with a filter media. The media is either in suspension or supported by a gravel layer at the foot of the filter. The dual purpose of this media is to support highly active biomass that is attached to it and to filter suspended solids. Carbon reduction and ammonia conversion occurs in aerobic mode and sometime achieved in a single reactor while nitrate conversion occurs in anoxic mode. BAF is operated either in upflow or downflow configuration depending on design specified by manufacturer.

Rotating biological contactors (RBCs) are mechanical secondary treatment systems, which are robust and capable of withstanding surges in organic load. RBCs were first installed in Germany in 1960 and have since been developed and refined into a reliable operating unit. The rotating disks support the growth of bacteria and micro-organisms present in the sewage, which break down and stabilise organic pollutants. To be successful, micro-organisms need both oxygen to live and food to grow. Oxygen is obtained from the atmosphere as the disks rotate. As the micro-organisms grow, they build up on the media until they are sloughed off due to shear forces provided by the rotating discs in the sewage. Effluent from the RBC is then passed through final clarifiers where the micro-organisms in suspension settle as a sludge. The sludge is withdrawn from the clarifier for further treatment.

A functionally similar biological filtering system has become popular as part of home aquarium filtration and purification. The aquarium water is drawn up out of the tank and then cascaded over a freely spinning corrugated fiber-mesh wheel before passing through a media filter and back into the aquarium. The spinning mesh wheel develops a biofilm coating of microorganisms that feed on the suspended wastes in the aquarium water and are also exposed to the atmosphere as the wheel rotates. This is especially good at removing waste urea and ammonia urinated into the aquarium water by the fish and other animals.

## Membrane Bioreactors

Membrane bioreactors (MBR) combine activated sludge treatment with a membrane liquid-solid separation process. The membrane component uses low pressure microfiltration or ultra filtration membranes and eliminates the need for clarification and tertiary filtration. The membranes are typically immersed in the aeration tank; however, some applications utilize a separate membrane tank. One of the key benefits of an MBR system is that it effectively overcomes the limitations associated with poor settling of sludge in conventional activated sludge (CAS) processes. The technology permits bioreactor operation with considerably higher mixed liquor suspended solids (MLSS) concentration than CAS systems, which are limited by sludge settling. The process is typically operated at MLSS in the range of 8,000-12,000 mg/L, while CAS are operated in the range of 2,000-3,000 mg/L. The elevated biomass concentration in the MBR process allows for very effective removal of both soluble and particulate biodegradable materials at higher loading rates. Thus increased sludge retention times, usually exceeding 15 days, ensure complete nitrification even in extremely cold weather.

The cost of building and operating an MBR is usually higher than conventional wastewater treatment. Membrane filters can be blinded with grease or abraded by suspended grit and lack a clarifier's flexibility to pass peak flows. The technology has become increasingly popular for reliably pretreated waste streams and has gained wider acceptance where infiltration and inflow have been controlled, however, and the life-cycle costs have been steadily decreasing. The small footprint of MBR systems, and the high quality effluent produced, make them particularly useful for water reuse applications.

### Secondary Sedimentation

The final step in the secondary treatment stage is to settle out the biological floc or filter material through a secondary clarifier and to produce sewage water containing low levels of organic material and suspended matter.

### Tertiary Treatment

The purpose of tertiary treatment is to provide a final treatment stage to raise the effluent quality before it is discharged to the receiving environment (sea, river, lake, ground, etc.). More than one tertiary treatment process may be used at any treatment plant. If disinfection is practiced, it is always the final process. It is also called 'effluent polishing'.

### Filtration

Sand filtration removes much of the residual suspended matter. Filtration over activated carbon, also called *carbon adsorption,* removes residual toxins.

Lagooning provides settlement and further biological improvement through storage in large man-made ponds or lagoons. These lagoons are highly aerobic and colonization by native macrophytes, especially reeds, is often encouraged. Small filter feeding invertebrates such as *Daphnia* and species of *Rotifera* greatly assist in treatment by removing fine particulates.

## Nutrient Removal

Wastewater may contain high levels of the nutrients nitrogen and phosphorus. Excessive release to the environment can lead to a build up of nutrients, called eutrophication, which can in turn encourage the overgrowth of weeds, algae, and cyanobacteria (blue-green algae). This may cause an algal bloom, a rapid growth in the population of algae. The algae numbers are unsustainable and eventually most of them die. The decomposition of the algae by bacteria uses up so much of oxygen in the water that most or all of the animals die, which creates more organic matter for the bacteria to decompose. In addition to causing deoxygenation, some algal species produce toxins that contaminate drinking water supplies. Different treatment processes are required to remove nitrogen and phosphorus.

## Nitrogen Removal

The removal of nitrogen is effected through the biological oxidation of nitrogen from ammonia to nitrate (nitrification), followed by denitrification, the reduction of nitrate to nitrogen gas. Nitrogen gas is released to the atmosphere and thus removed from the water.

Nitrification itself is a two-step aerobic process, each step facilitated by a different type of bacteria. The oxidation of ammonia ($NH_3$) to nitrite ($NO_2^-$) is most often facilitated by *Nitrosomonas* spp. (nitroso referring to the formation of a nitroso functional group). Nitrite oxidation to nitrate ($NO_3^-$), though traditionally believed to be facilitated by *Nitrobacter* spp. (nitro referring the formation of a nitro functional group), is now known to be facilitated in the environment almost exclusively by *Nitrospira* spp. Denitrification requires anoxic conditions to encourage the appropriate biological communities to form. It is facilitated by a wide diversity of bacteria. Sand filters, lagooning and reed beds can all be used to reduce nitrogen, but the activated sludge process (if designed well) can do the job the most easily. Since denitrification is the reduction of nitrate to dinitrogen gas, an electron donor is needed. This can be, depending on the wastewater, organic matter (from faeces), sulfide, or an added donor like methanol.

Sometimes the conversion of toxic ammonia to nitrate alone is referred to as tertiary treatment.

Many sewage treatment plants use axial flow pumps to transfer the nitrified mixed liquor from the aeration zone to the anoxic zone for denitrification. These pumps are often referred to as *Internal Mixed Liquor Recycle* (IMLR) pumps. The sludge in the anoxic tanks must be mixed well (mixture of recirculated mixed liquor, return activated sludge [RAS], and raw influent) by using submersible mixers in order to achieve the desired denitrification.

## Phosphorus Removal

Phosphorus removal is important as it is a limiting nutrient for algae growth in many fresh water systems. (For a description of the negative effects of algae. It is also particularly important for water reuse systems where high phosphorus concentrations may lead to fouling of downstream equipment such as reverse osmosis.

Phosphorus can be removed biologically in a process called enhanced biological phosphorus removal. In this process, specific bacteria, called polyphosphate accumulating organisms (PAOs), are selectively enriched and accumulate large quantities of phosphorus within their cells (up to 20 per cent of their mass). When the biomass enriched in these bacteria is separated from the treated water, these biosolids have a high fertilizer value.

Phosphorus removal can also be achieved by chemical precipitation, usually with salts of iron (e.g. ferric chloride), aluminum (e.g. alum), or lime. This may lead to excessive sludge production as hydroxides precipitates and the added chemicals can be expensive. Chemical phosphorus removal requires significantly smaller equipment footprint than biological removal, is easier to operate and is often more reliable than biological phosphorus removal. Another method for phosphorus removal is to use granular laterite.

Once removed, phosphorus, in the form of a phosphate-rich sludge, may be stored in a land fill or resold for use in fertilizer.

## Disinfection

The purpose of disinfection in the treatment of wastewater is to substantially reduce the number of microorganisms in the water to be discharged back into the environment. The effectiveness of disinfection depends on the quality of the water being treated (e.g., cloudiness, pH, etc.), the type of disinfection being used, the disinfectant dosage (concentration and time), and other environmental variables. Cloudy water will be treated less successfully, since solid matter can shield organisms, especially from ultraviolet light or if contact times are low. Generally, short contact times, low doses and high flows all militate against effective disinfection. Common methods of disinfection include ozone, chlorine, ultraviolet light, or sodium hypochlorite. Chloramine, which is used for drinking water, is not used in wastewater treatment because of its persistence.

Chlorination remains the most common form of wastewater disinfection in North America due to its low cost and long-term history of effectiveness. One disadvantage is that chlorination of residual organic material can generate chlorinated-organic compounds that may be carcinogenic or harmful to the environment. Residual chlorine or chloramines may also be capable of chlorinating organic material in the natural aquatic environment. Further, because residual chlorine is toxic to aquatic species, the treated effluent must also be chemically dechlorinated, adding to the complexity and cost of treatment.

Ultraviolet (UV) light can be used instead of chlorine, iodine, or other chemicals. Because no chemicals are used, the treated water has no adverse effect on organisms that later consume it, as may be the case with other methods. UV radiation causes damage to the genetic structure of bacteria, viruses, and other pathogens, making them incapable of reproduction. The key disadvantages of UV disinfection are the need for frequent lamp maintenance and replacement and the need for a highly treated effluent to ensure that the target microorganisms are not shielded from the UV radiation (i.e., any solids present in the treated effluent may protect microorganisms from the UV light). In the United Kingdom, UV light is becoming the most common means of disinfection because of the concerns about the impacts of chlorine in chlorinating residual organics in the wastewater and in chlorinating organics in the receiving water. Some sewage treatment systems in Canada and the US also use UV light for their effluent water disinfection.

Ozone ($O_3$) is generated by passing oxygen ($O_2$) through a high voltage potential resulting in a third oxygen atom becoming attached and forming $O_3$. Ozone is very unstable and reactive and oxidizes most organic material it comes in contact with, thereby destroying many pathogenic microorganisms. Ozone is considered to be safer than chlorine because, unlike chlorine which has to be stored on site (highly poisonous in the event of an accidental release), ozone is generated onsite as needed. Ozonation also produces fewer disinfection by-products than chlorination. A disadvantage of ozone disinfection is the high cost of the ozone generation equipment and the requirements for special operators.

## Odour Control

Odours emitted by sewage treatment are typically an indication of an anaerobic or 'septic' condition. Early stages of processing will tend to produce smelly gases, with hydrogen sulfide being most common in generating complaints. Large process plants in urban areas will often treat the odours with carbon reactors, a contact media with bio-slimes, small doses of chlorine, or circulating fluids to biologically capture and metabolize the obnoxious gases. Other methods of odour control exist, including addition of iron salts, hydrogen peroxide, calcium nitrate, etc. to manage hydrogen sulfide levels.

## Package Plants and Batch Reactors

To use less space, treat difficult waste and intermittent flows, a number of designs of hybrid treatment plants have been produced. Such plants often combine at least two stages of the three main treatment stages into one combined stage. In the UK, where a large number of wastewater treatment plants serve small populations, package plants are a viable alternative to building a large structure for each process stage. In the US, package plants are typically used in rural areas, highway rest stops and trailer parks.

One type of system that combines secondary treatment and settlement is the sequencing batch reactor (SBR). Typically, activated sludge is mixed with raw incoming sewage, and then mixed and aerated. The settled sludge is run off and re-aerated before a proportion is returned to the headworks. SBR plants are now being deployed in many parts of the world.

The disadvantage of the SBR process is that it requires a precise control of timing, mixing and aeration. This precision is typically achieved with computer controls linked to sensors. Such a complex, fragile system is unsuited to places where controls may be unreliable, poorly maintained, or where the power supply may be intermittent.

Package plants may be referred to as *high charged* or *low charged*. This refers to the way the biological load is processed. In high charged systems, the biological stage is presented with a high organic load and the combined floc and organic material is then oxygenated for a few hours before being charged again with a new load. In the low charged system the biological stage contains a low organic load and is combined with flocculate for longer times.

## Sludge Treatment and Disposal

The sludges accumulated in a wastewater treatment process must be treated and disposed of in a safe and effective manner. The purpose of digestion is to reduce the amount of organic matter and the number of disease-causing microorganisms present in the solids. The most common treatment options include anaerobic digestion, aerobic digestion, and composting. Incineration is also used albeit to a much lesser degree.

Sludge treatment depends on the amount of solids generated and other site-specific conditions. Composting is most often applied to small-scale plants with aerobic digestion for mid sized operations, and anaerobic digestion for the larger-scale operations.

## Anaerobic Digestion

Anaerobic digestion is a bacterial process that is carried out in the absence of oxygen. The process can either be *thermophilic* digestion, in which sludge is fermented in tanks at a temperature of 55° C, or *mesophilic*, at a temperature of around 36° C. Though allowing shorter retention time (and thus smaller tanks), thermophilic digestion is more expensive in terms of energy consumption for heating the sludge.

Anaerobic digestion is the most common (mesophilic) treatment of domestic sewage in septic tanks, which normally retain the sewage from one day to two days, reducing the BOD by about 35 to 40 per cent. This reduction can be increased with a combination of anaerobic and aerobic treatment by installing *Aerobic Treatment Units* (ATUs) in the septic tank.

One major feature of anaerobic digestion is the production of biogas (with the most useful component being methane), which can be used in generators for electricity production and/or in boilers for heating purposes.

## Aerobic Digestion

Aerobic digestion is a bacterial process occurring in the presence of oxygen. Under aerobic conditions, bacteria rapidly consume organic matter and convert it into carbon dioxide. The operating costs used to be characteristically much greater for aerobic digestion because of the energy used by the blowers, pumps and motors needed to add oxygen to the process.

Aerobic digestion can also be achieved by using diffuser systems or jet aerators to oxidize the sludge.

## Composting

Composting is also an aerobic process that involves mixing the sludge with sources of carbon such as sawdust, straw or wood chips. In the presence of oxygen, bacteria digest both the wastewater solids and the added carbon source and, in doing so, produce a large amount of heat.

## Incineration

Incineration of sludge is less common because of air emissions concerns and the supplemental fuel (typically natural gases or fuel oil) required to burn the low calorific value sludge and vaporize residual water. Stepped multiple hearth incinerators with high residence time and fluidized bed incinerators are the most common systems used to combust wastewater sludge. Co-firing in municipal waste-to-energy plants is occasionally done, this option being less expensive assuming the facilities already exist for solid waste and there is no need for auxiliary fuel.

## Sludge Disposal

When a liquid sludge is produced, further treatment may be required to make it suitable for final disposal. Typically, sludges are thickened (dewatered) to reduce the volumes transported off-site for disposal. There is no process which completely eliminates the need to dispose of biosolids. There is, however, an additional step some cities are taking to superheat sludge and convert it into small pelletized granules that are high in nitrogen and other organic materials. In New York City, for example, several sewage treatment plants have dewatering facilities that use large centrifuges along with the addition of chemicals such as polymer to further remove liquid from the sludge. The removed fluid, called centrate, is typically reintroduced into the wastewater process. The product which is left is called 'cake' and that is picked up by companies which turn it into fertilizer pellets. This product is then sold to local farmers and turf farms as a soil amendment or fertilizer, reducing the amount of space required to dispose of sludge in landfills. Much sludge originating from commercial or industrial areas is contaminated with toxic materials that are released into the sewers from the industrial processes Elevated concentrations of such materials may make the sludge unsuitable for agricultural use and it may then have to be incinerated or disposed of to landfill.

Many processes in a wastewater treatment plant are designed to mimic the natural treatment processes that occur in the environment, whether that environment is a natural water body or the ground. If not overloaded, bacteria in the environment will consume organic contaminants, although this will reduce the levels of oxygen in the water and may significantly change the overall ecology of the receiving water. Native bacterial populations feed on the organic contaminants, and the numbers of disease-causing microorganisms are reduced by natural environmental conditions such as predation or exposure to ultraviolet radiation. Consequently, in cases where the receiving environment provides a high level of dilution, a high degree of wastewater treatment may not be required. However, recent evidence has demonstrated that very low levels of specific contaminants in wastewater, including hormones (from animal husbandry and residue from human hormonal contraception methods) and synthetic materials such as phthalates that mimic hormones in their action, can have an unpredictable adverse impact on the natural biota and potentially on humans if the water is re-used for drinking water. In the US and EU, uncontrolled discharges of wastewater to the environment are not permitted under law, and strict water quality requirements are to be met. (For requirements in the US, A significant threat in the coming decades will be the increasing uncontrolled discharges of wastewater within rapidly developing countries.

## Effects on Biology

Sewage treatment plants can have multiple effects on nutrient levels in the water that the treated sewage flows into. These effects on nutrients can have large effects on the biological life in the water in contact with the effluent. Treatment ponds can include any of the following:

- Oxidation ponds, which are aerobic bodies of water usually 1-2 meters in depth that receive effluent from sedimentation tanks or other forms of primary treatment.
- Dominated by algae
- Polishing ponds are similar to oxidation ponds but receive effluent from an oxidation pond or from a plant with an extended mechanical treatment.
- Dominated by zooplankton
- Raw sewage lagoons or sewage lagoons are aerobic ponds where sewage is added with no primary treatment other than coarse screening.
- Anaerobic lagoons are heavily loaded ponds.
- Dominated by bacteria.
- Sludge lagoons are aerobic ponds, usually 2-5 meters in depth, that receive anaerobically digested primary sludge, or activated secondary sludge under water.
- Upper layers are dominated by algae.

Phosphorous limitation is a possible result from sewage treatment and results in flagellate-dominated plankton, particularly in summer and fall.

At the same time a different study found high nutrient concentrations linked to sewage effluents. High nutrient concentration leads to high chlorophyll a concentrations, which is a proxy for primary production in marine environments. High primary production means high phytoplankton populations and most likely high zooplankton populations because zooplankton feed on phytoplankton. However, effluent released into marine systems also leads to greater population instability.

A study done in Britain found that the quality of effluent effected the planktonic life in the water in direct contact with the wastewater effluent. Turbid, low-quality effluents either did not contain ciliated protozoa or contained only a few species in small numbers. On the other hand, high-quality effluents contained a wide variety of ciliated protozoa in large numbers. Due to these findings, it seems unlikely that any particular component of the industrial effluent has, by itself, any harmful effects on the protozoan populations of activated sludge plants.

The planktonic trends of high populations close to input of treated sewage is contrasted by the bacterial trend. In a study of *Aeromonas* spp. in increasing distance from a wastewater source, greater change in seasonal cycles was found the furthest from the effluent. This trend is so strong that the furthest location studied actually had an inversion of the *Aeromonas* spp. cycle in comparison to that of fecal coliforms. Since there is a main pattern in the cycles that occurred simultaneously at all stations it indicates seasonal factors (temperature, solar radiation, phytoplankton) control of the bacterial population. The effluent dominant species changes from *Aeromonas caviae* in winter to *Aeromonas sobria* in the spring and fall while the inflow dominant species is *Aeromonas caviae*, which is constant throughout the seasons.

## Sewage Treatment in Developing Countries

Few reliable figures on the share of the wastewater collected in sewers that is being treated in the world exist. In many developing countries the bulk of domestic and industrial wastewater is discharged without any treatment or after primary treatment only. In Latin America about 15 per cent of collected wastewater passes through treatment plants (with varying levels of actual treatment). In Venezuela, a below average country in South America with respect to wastewater treatment, 97 per cent of the country's sewage is discharged raw into the environment. In a relatively developed Middle Eastern country such as Iran, Tehran's majority of population has totally untreated sewage injected to the city's groundwater.

In Israel, about 50 per cent of agricultural water usage (total use was 1 billion cubic metres in 2008) is provided through reclaimed sewer water. Future plans call for increased use of treated sewer water as well as more desalination plants.

# Wastewater

Wastewater is any water that has been adversely affected in quality by anthropogenic influence. It comprises liquid waste discharged by domestic residences, commercial properties, industry, and/or agriculture and can encompass a wide range of potential contaminants and concentrations. In the most common usage, it refers to the municipal wastewater that contains a broad spectrum of contaminants resulting from the mixing of wastewaters from different sources.

Sewage is correctly the subset of wastewater that is contaminated with feces or urine, but is often used to mean any waste water. 'Sewage' includes domestic, municipal, or industrial liquid waste products disposed of, usually via a pipe or sewer or similar structure, sometimes in a cesspool emptier.

The physical infrastructure, including pipes, pumps, screens, channels etc. used to convey sewage from its origin to the point of eventual treatment or disposal is termed sewerage.

Wastewater or sewage can come from (text in brackets indicates likely inclusions or contaminants):

- Human waste ( fæces, used toilet paper or wipes, urine, or other bodily fluids), also known as blackwater, usually from lavatories;
- Cesspit leakage;
- Septic tank discharge;
- Sewage treatment plant discharge;
- Washing water (personal, clothes, floors, dishes, etc.), also known as greywater or sullage;

- Rainfall collected on roofs, yards, hard-standings, etc. (generally clean with traces of oils and fuel);
- Groundwater infiltrated into sewage;
- Surplus manufactured liquids from domestic sources (drinks, cooking oil, pesticides, lubricating oil, paint, cleaning liquids, etc.);
- Urban rainfall runoff from roads, carparks, roofs, sidewalks, or pavements (contains oils, animal fæces, litter, fuel or rubber residues, metals from vehicle exhausts, etc.);
- Seawater ingress (high volumes of salt and micro-biota);
- Direct ingress of river water (high volumes of micro-biota);
- Direct ingress of manmade liquids (illegal disposal of pesticides, used oils, etc.);
- Highway drainage (oil, de-icing agents, rubber residues);
- Storm drains (almost anything, including cars, shopping trolleys, trees, cattle, etc.);
- Blackwater (surface water contaminated by sewage);
- Industrial waste;
- industrial site drainage (silt, sand, alkali, oil, chemical residues);
- Industrial cooling waters (biocides, heat, slimes, silt);
- Industrial process waters;
- Organic or bio-degradable waste, including waste from abattoirs, creameries, and ice cream manufacture;
- Organic or non-bio degradable/difficult-to-treat waste (pharmaceutical or pesticide manufacturing);
- extreme pH waste (from acid/alkali manufacturing, metal plating);
- Toxic waste (metal plating, cyanide production, pesticide manufacturing, etc.);
- Solids and Emulsions ( paper manufacturing, foodstuffs, lubricating and hydraulic oil manufacturing, etc.);
- agricultural drainage, direct and diffuse.

## Wastewater Constituents

The composition of wastewater varies widely. This is a partial list of what it may contain:

- Water (> 95%) which is often added during flushing to carry waste down a drain;

- Pathogens such as bacteria, viruses, prions and parasitic worms;
- Non-pathogenic bacteria;
- Organic particles such as faeces, hairs, food, vomit, paper fibres, plant material, humus, etc.;
- Soluble organic material such as urea, fruit sugars, soluble proteins, drugs, pharmaceuticals, etc.;
- Inorganic particles such as sand, grit, metal particles, ceramics, etc.;
- Soluble inorganic material such as ammonia, road-salt, sea-salt, cyanide, hydrogen sulfide, thiocyanates, thiosulfates, etc.;
- Animals such as protozoa, insects, arthropods, small fish, etc.;
- Macro-solids such as sanitary napkins, nappies/diapers, condoms, needles, children's toys, dead animals or plants, etc.;
- Gases such as hydrogen sulfide, carbon dioxide, methane, etc.;
- Emulsions such as paints, adhesives, mayonnaise, hair colorants, emulsified oils, etc.;
- Toxins such as pesticides, poisons, herbicides, etc.

Any oxidizable material present in a natural waterway or in an industrial wastewater will be oxidized both by biochemical (bacterial) or chemical processes. The result is that the oxygen content of the water will be decreased. Basically, the reaction for biochemical oxidation may be written as:

Oxidizable material + bacteria + nutrient + $O_2$=$CO_2$ + $H_2O$ + oxidized inorganics such as $NO_3$ or $SO_4$

Oxygen consumption by reducing chemicals such as sulfides and nitrites is typified as follows:

$S^- + 2\,O_2 = SO_4^-$

$NO_2^- + \frac{1}{2}\,O_2 = NO_3^-$

Since all natural waterways contain bacteria and nutrients, almost any waste compounds introduced into such waterways will initiate biochemical reactions (such as shown above). Those biochemical reactions create what is measured in the laboratory as the Biochemical oxygen demand (BOD). Such chemicals are also liable to be broken down using strong oxidising agents and these chemical reactions create what is measured in the laboratory as the Chemical oxygen demand (COD).

Both the BOD and COD tests are a measure of the relative oxygen-depletion effect of a waste contaminant. Both have been widely adopted as a measure of pollution effect. The BOD test measures the oxygen demand of biodegradable pollutants whereas the COD test measures the oxygen demand of oxidizable pollutants.

The so-called 5-day BOD measures the amount of oxygen consumed by biochemical oxidation of waste contaminants in a 5-day period. The total amount of oxygen consumed when the biochemical reaction is allowed to proceed to completion is called the Ultimate BOD. The Ultimate BOD is too time consuming, so the 5-day BOD has almost universally been adopted as a measure of relative pollution effect.

There are also many different COD tests of which the 4-hour COD is probably the most common.

There is no generalized correlation between the 5-day BOD and the ultimate BOD. Similarly there is no generalized correlation between BOD and COD. It is possible to develop such correlations for a specific waste contaminants in a specific waste water stream but such correlations cannot be generalized for use with any other waste contaminants or waste water streams. This is because the composition of any waste water stream is different. As an example and effluent consisting of a solution of simple sugars that might discharge from a confectionery factory is likely to have organic components that degrade very quickly. In such a case the 5 day BOD and the ultimate BOD would be very similar, i.e there would be very little organic material left after 5 days. However a final effluent of a sewage treatment works serving a large industrialised area might have a discharge where the ultimate BOD was much greater than the 5 day BOD because much of the easily degraded material would have been removed in the sewage treatment process and many industrial processes discharge difficult to degrade organic molecules.

The laboratory test procedures for the determining the above oxygen demands are detailed in many standard texts. American versions include the 'Standard Methods For the Examination Of Water and Wastewater'.

In some urban areas, sewage is carried separately in sanitary sewers and runoff from streets is carried in storm drains. Access to either of these is typically through a manhole. During high precipitation periods a sanitary sewer overflow can occur, causing potential public health and ecological damage.

Sewage may drain directly into major watersheds with minimal or no treatment. When untreated, sewage can have serious impacts on the quality of an environment and on the health of people. Pathogens can cause a variety of illnesses. Some chemicals pose risks even at very low concentrations and can remain a threat for long periods of time because of bioaccumulation in animal or human tissue.

There are numerous processes that can be used to clean up waste waters depending on the type and extent of contamination. Most wastewater is treated in industrial-scale wastewater treatment plants (WWTPs) which may include physical, chemical and biological treatment processes. However, the use of septic tanks and other On-Site Sewage Facilities (OSSF) is widespread in rural areas, serving up to one quarter of the homes in the U.S. The most important aerobic treatment system is the activated sludge process, based on the maintenance and recirculation of a complex biomass composed by micro-organisms able to absorb and adsorb the organic matter carried in the wastewater. Anaerobic processes are widely applied in the treatment of industrial wastewaters and biological sludge. Some wastewater may be highly treated and reused as reclaimed water. For some waste waters ecological approaches using reed bed systems such as constructed wetlands may be appropriate. Modern systems include tertiary treatment by micro filtration or synthetic membranes. After membrane filtration, the treated wastewater is indistinguishable from waters of natural origin of drinking quality. Nitrates can be removed from wastewater by microbial denitrification, for which a small amount of methanol is typically added to provide the bacteria with a source of carbon. Ozone Waste Water Treatment is also growing in popularity, and requires the use of an ozone generator, which decontaminates the water as Ozone bubbles percolate through the tank.

Disposal of wastewaters from an industrial plant is a difficult and costly problem. Most petroleum refineries, chemical and petrochemical plants have onsite facilities to treat their wastewaters so that the pollutant concentrations in the treated wastewater comply with the local and/or national regulations regarding disposal of wastewaters into community treatment plants or into rivers, lakes or oceans. Other Industrial processes that produce a lot of waste-waters such as paper and pulp production has created environmental concern leading to development of processes to recycle water use within plants before they have to be cleaned and disposed of.

## Reuse

Treated wastewater can be reused as drinking water, in industry (cooling towers), in artificial recharge of aquifers, in agriculture (70% of Israel's irrigated agriculture is based on highly purified wastewater) and in the rehabilitation of natural ecosystems (Florida's Everglades).

## Algal Fuel

Woods Hole Oceanographic Institution and Harbor Branch Oceanographic Institution, following the conclusions of the USDOE´s Aquatic Species Programme, use wastewater for breeding algae. The wastewater from domestic and industrial sources contain rich organic compounds, which accelerate the growth of algae. This algae can be used to produce algal fuels.

Algaewheel, based in Indianapolis, Indiana, presented a proposal to build a new wastewater treatment facility in Cedar Lake, Indiana that uses algae to treat municipal wastewater and uses the sludge byproduct to produce biofuel.

### Etymology

The words 'sewage' and 'sewer' came from Old French *essouier* = 'to drain', which came from Latin *exaquare*. Their formal Latin antecedents are *exaquaticum* and *exaquarium*.

### Municipal Solid Waste

Municipal Solid Waste (MSW), also called urban solid waste, is a waste type that includes predominantly household waste (domestic waste) with sometimes the addition of commercial wastes collected by a municipality within a given area. They are in either solid or semisolid form and generally exclude industrial hazardous wastes. The term *residual waste* relates to waste left from household sources containing materials that have not been separated out or sent for reprocessing.

- *Biodegradable waste:* food and kitchen waste, green waste, paper (can also be recycled).
- *Recyclable material:* paper, glass, bottles, cans, metals, certain plastics, etc.
- *Inert waste:* construction and demolition waste, dirt, rocks, debris.
- *Composite wastes:* waste clothing, Tetra Paks, waste plastics such as toys.
- *Domestic hazardous waste* (also called 'household hazardous waste') and toxic waste: medication, e-waste, paints, chemicals, light bulbs, fluorescent tubes, spray cans, fertilizer and pesticide containers, batteries, shoe polish.

## THE FUNCTIONAL ELEMENTS OF SOLID WASTE

### Waste Generation

Waste generation encompasses activities in which materials are identified as no longer being of value and are either thrown out or gathered together for disposal.

### Waste Handling and Separation, Storage and Processing at the Source

Waste handling and separation involves the activities associated with management of waste until they are placed in storage container for collection. Handling also encompasses the movement of loaded containers to the point of collection. Separation of waste components is an important step in the handling and storage of solid waste at the source.

### Collection

The functional element of collection includes not only the gathering of solid waste and recyclable materials, but also the transport of these materials, after collection, to the location where the collection vehicle is emptied. This location may be a materials processing facility, a transfer station or a landfill disposal site.

### Separation and Processing and Transformation of Solid Wastes

The types of means and facilities that are now used for the recovery of waste materials that have been separated at the source include curbside collection, drop off and buy back centres. The separation and processing of wastes that have been separated at the source and the separation of commingled wastes usually occur at a materials recovery facility, transfer stations, combustion facilities and disposal sites.

### Transfer and Transport

This element involves two steps:

1. the transfer of wastes from the smaller collection vehicle to the larger transport equipment;
2. the subsequent transport of the wastes, usually over long distances, to a processing or disposal site.

### Disposal

Today the disposal of wastes by land filling or land spreading is the ultimate fate of all solid wastes, whether they are residential wastes collected and transported directly to a landfill site, residual materials from materials recovery facilities (MRFs), residue from the combustion of solid waste, compost or other substances from various solid waste processing facilities. A modern sanitary landfill is not a dump; it is an engineered facility used for disposing of solid wastes on land without creating nuisances or hazards to public health or safety, such as the breeding of insects and the contamination of ground water.

## Energy Generation

Municipal solid waste can be used to generate energy. Several technologies have been developed that make the processing of MSW for energy generation cleaner and more economical than ever before, including landfill gas capture, combustion, pyrolysis, gasification, and plasma arc gasification. While older waste incineration plants emitted high levels of pollutants, recent regulatory changes and new technologies have significantly reduced this concern. EPA regulations in 1995 and 2000 under the Clean Air Act have succeeded in reducing emissions of dioxins from waste-to-energy facilities by more than 99 per cent below 1990 levels, while mercury emissions have been by over 90 per cent. The EPA noted these improvements in 2003, citing waste-to-energy as a power source "with less environmental impact than almost any other source of electricity".

## Litter

Litter consists of waste products such as containers, papers, and wrappers which have been disposed of without consent. Litter can also be used as a verb. To litter means to throw litter onto the ground as opposed to disposing of it properly.

Throughout human history, people have disposed of unwanted materials onto streets, countrysides and remote places, unpunished. Prior to reforms within cities in the mid-to-late 19th century, sanitation was not a priority on governments' lists of things to do. Waste was disposed of by the roadside or in small local dumps. It was unsanitary for local inhabitants and the growing piles of waste led to the spread of disease. The only known pre-modern exception, however, was the Arab Empire, especially in Cordoba, al-Andalus, which had facilities for litter collection.

In the 14th century, the rise of waste in Europe helped contribute to the bubonic plague. Black rats, carrying the fleas which were the vectors for the plague, fed off the food scraps. Many plagues were started from litter.

In addition to intentional littering, almost half of litter on U.S. roadways is now a result of accidental or unintentional litter, debris that falls off of improperly secured trash and recycling collection vehicles and pickup trucks. Heavy traffic and proximity to waste disposal sites are known to correlate with higher litter rates.

According to a study by the Dutch organisation VROM, 80 per cent of the people claim that 'everybody leaves of a piece of paper, tin or something, on the street behind'. Young people from 12 to 24 years cause more litter than the average (Dutch or Belgian) person. 18 per cent of people who regularly cause litter were 50 years of age or older. Nevertheless, automobile drivers and recreationalists, smokers and the youth are specific target groups within many campaigns conducted to keep countries free of litter.

**Effects**

Litter can harm the environment in a number of different ways. It is a breeding ground for disease-causing insects and rodents. Its 'ugliness' damages the appearance of scenic environments. Open containers such as paper cups or beverage cans can hold rainwater, providing breeding locations for mosquitoes which have been known to spread diseases such as West Nile Virus and Malaria. Uncollected litter can attract more, flowing into streams, and storm water drainage systems, local bays and estuaries. About 18 per cent of litter, usually traveling through storm water systems, ends up in local streams, rivers, and waterways. About 80 per cent of marine debris comes from land-based sources. Animals may get trapped or poisoned with litter in their habitats. Cigarette butts and filters are a threat to wildlife and have been found in the stomachs of fish, birds and whales, who have mistaken them for food. Debris falling from vehicles is an increasing cause of automobile accidents. Cleaning up litter in the U.S. costs hundreds of dollars per ton, about ten times more than the cost of trash disposal, for a cost totaling about $11 billion per year.

Litter can be expensive to clean up, so the act of littering has been made a fineable offense by statute in many places.

Some jurisdictions offer small bounties for preventing and cleaning up litter (for example, requiring people to pay a deposit on bottles, which is only given when the bottles are returned).

**Australia**

Litter in Australia is prevalent in many areas. An anti-litter movement began in 1969 in Victoria with the formation of Keep Australia Beautiful. Its major anti-littering campaigns 'Do the right thing' and 'Tidy Towns' became well known nationally. Today, the most vocal organisation is Clean Up Australia which holds a national clean up day. There is currently no Government of Australia legislation against litter. Legislation is generally considered the responsibility of either an States and territories of Australia (Environmental Protection Agency) or Local Government Areas. All states and territories now have legislation against littering which may include fines, which are enforceable by the police or other agents. The first state level anti-litter legislation in Australia was the Environment Protection Act (1970) introduced in Victoria. Some state environmental protection agencies (including Victoria and Queensland) do online litter reports.

### The Netherlands

In the Netherlands, the County Board municipalities as well as a special division called Rijkswaterstaat play a role in cleaning up litter. The Dutch police and local supervisors (called buitengewoon opsporingsambtenaar, or BOA's) fine citizens for throwing away cans, bottles or wrappers onto the street. The fine is 90 euro, unless the defendant is between 12 and 16 years old (the penalty is then half the amount, or 45 euro).

### New Zealand

Litter in New Zealand is prevalent in many areas, such as streets, waterways, drainage ditches, forests, and beaches. The Litter Act was adopted in 1979; it established the Keep New Zealand Beautiful organisation.

### Singapore

Litter in Singapore has very rare and unusual restrictions. Singapore is known for its strict laws against littering of any kind. In order to maintain their reputation as the cleanest city in the world, fines of S$1,000 for dropping a simple piece of trash are typical. The Corrective Work Order requires repeat offenders to spend a few hours cleaning a public place wearing bright jackets, and in some cases, the local media are invited to cover the public spectacle. Despite all the attempts, tourists can still see Singaporeans spitting and littering in food courts. Chinese males in middle age are likely to spit bones directly on the tables in public food courts and leave the place unwiped.

### United Kingdom

Littering is an offence under S87 of the Environmental Protection Act 1990. S88 allows authorised officers to issue a Fixed Penalty Notice (FPN) of usually £80 (though this can vary from authority to authority, a lot of Councils now have some sort of wardens patrolling to issue for these kinds of offences) to allow offenders to discharge liability for this offence in a relatively simple way. They can also be prosecuted, the maximum fine in the Magistrates Court is £2500, they can also go to Court if they are offered an FPN but want a trial as is an individuals right in the UK.

### United States

Litter in the United States is an environmental issue, and littering is often an offense punishable with a fine as set out by statutes in many places. Litter laws, enforcement efforts, and court prosecutions are used to help curtail littering.

The American Public Works Association standardized the term litter in the mid-20th century, to be later known as a form of solid waste—"material which, if thrown or deposited, tends to create a danger to public health, safety and welfare." Litter is categorized into three specific components: hazardous, reusable-recyclable and non-hazardous, non-from trash-hauling vehicles, unsecured loads, or construction sites.

## International Waters

In international waters, great amounts of litter have been found floating around. According to UNEP, 6.4 million tons of litter wind up in the oceans per year of which a large part are drawn into international waters because of sea currents. An example of this is the Plastic Vortex. Litter from all over the world flows into this one central area because of the sea currents.

In the Pacific Ocean and near the shores of the Pacific Islands, two piles of waste drive around, which contain both around 100 million tons of litter. Most of the litter is situated in the North Pacific Subtropical Convergence Zone and is referred to as the 'Plastic Vortex'. The litter consists primarily of plastic, according to scientists, from lego blocks and footballs to broken kayaks. Scientists are already fully aware of the existence of the floating litter, which was discovered accidentally in 1997, but since then has increased heavily in size.

CHAPTER 8

# Hazardous Waste

A hazardous waste is waste that poses substantial or potential threats to public health or the environment. There are four factors that determine whether or not a substance is hazardous:

- ignitability (i.e., flammable)
- reactivity
- corrosivity
- toxicity

U.S. environmental laws additionally describe a 'hazardous waste' as a waste (usually a solid waste) that has the potential to:

- cause, or significantly contribute to an increase in mortality (death) or an increase in serious irreversible, or incapacitating reversible illness; or
- pose a substantial (present or potential) hazard to human health or the environment when improperly treated, stored, transported, or disposed of, or otherwise managed.

These wastes may be found in different physical states such as gaseous, liquids, or solids. Furthermore, a hazardous waste is a special type of waste because it cannot be disposed of by common means like other by-products of our everyday lives. Depending on the physical state of the waste, treatment and solidification processes might be available. In other cases, however, there is not much that can be done to prevent harm.

## Resource Conservation and Recovery Act (RCRA)

Modern hazardous waste regulations in the U.S. began with the Resource Conservation and Recovery Act (RCRA) which was enacted in 1976. The primary contribution of RCRA was to create a 'cradle to grave' system of record keeping for hazardous wastes. Hazardous wastes must be tracked from the time they are generated until their final disposition.

RCRA's record keeping system helps to track the life cycle of hazardous waste and reduces the amount of hazardous waste illegally disposed.

## Comprehensive Environmental Response, Compensation, and Liability Act

The Comprehensive Environmental Response, Compensation, and Liability Act (CERCLA), was enacted in 1980. The primary contribution of CERCLA was to create a 'Superfund' and provided for the clean-up and remediation of closed and abandoned hazardous waste sites.Before this act was created, hazardous wastes were being disposed in regular landfills until scientists measured unfavorable amounts of hazardous materials seeping into the ground. These chemicals eventually made their way to the water systems, and contaminated the soil that animals and crops used, as well as the soil that people employed to build their communities. After these regulations were put into practice, many landfills require now countermeasures against groundwater contamination; for example installing a barrier along the foundation of the landfill to contain the hazardous substances that may remain in the disposed waste. Currently, in order to enter a landfill, hazardous wastes must be stabilized and solidified, thus the new waste produced is less harmful than the original.

## Hazardous Wastes in the United States

Many types of businesses generate hazardous waste. Some are small areas that may be located in a community. For example, dry cleaners, automobile repair shops, hospitals, exterminators, and photo processing centers all generate hazardous waste. Some hazardous waste generators are larger companies like chemical manufacturers, electroplating companies, and oil refineries.

A US facility that treats, stores or disposes of hazardous waste must obtain a permit for doing so under the Resource Conservation and Recovery Act. Generators of and transporters of hazardous waste must meet specific requirements for handling, managing, and tracking waste. Through the RCRA, Congress directed the United States Environmental Protection Agency (EPA) to create regulations to manage hazardous waste. Under this mandate, the EPA developed strict requirements for all aspects of hazardous waste management including the treatment, storage, and disposal of hazardous waste. In addition to these federal requirements, states may develop more stringent requirements or requirements that are broader in scope than the federal regulations.

In the United States, hazardous wastes generated by commercial or industrial activities may be classified as 'listed' hazardous wastes or 'characteristic' hazardous wastes by the EPA.

In regulatory terms, a Resource Conservation and Recovery Act (RCRA) hazardous waste is a waste that either a 'characteristic waste' or a 'listed waste':

- *Characteristic Waste* — exhibits at least one of the four "characteristics" of hazardous waste (ignitability, corrosivity, reactivity, or toxicity)
- *Listed Waste* — appears on one of the four hazardous wastes lists (F-list, K-list, P-list, or U-list), or

Individual states may regulate particular wastes more stringently than mandated by federal regulation. This is because the U.S. EPA is authorized to delegate primary rulemaking authorization to individual states. Most states take advantage of this authority, implementing their own hazardous waste programmes that are at least as stringent as the federal programme.

**Characteristic Wastes**

Characteristic Hazardous Wastes are defined as wastes that exhibit the following characteristics: ignitability, corrosivity, reactivity, or toxicity.

**Ignitability**

Ignitable wastes can create fires under certain conditions, are spontaneously combustible, or have a flash point less than 60° C (140° F). Examples include waste oils and used solvents. For more details, see 40 CFR §261.21. Test methods that may be used to determine ignitability include the Pensky-Martens Closed-Cup Method for Determining Ignitability, the Setaflash Closed-Cup Method for Determining Ignitability, and the Ignitability of Solids.

**Corrosive**

Corrosive wastes are acids or bases (pH less than or equal to 2, or greater than or equal to 12.5) that are capable of corroding metal containers, such as storage tanks, drums, and barrels. Battery acid is an example. For more details, see 40 CFR §261.22. The test method that may be used to determine corrosivity is the Corrosivity Towards Steel.

**Reactivity**

Reactive wastes are unstable under 'normal' conditions. They can cause explosions, toxic fumes, gases, or vapors when heated, compressed, or mixed with water. Examples include lithium-sulfur batteries and explosives. For more details, see 40 CFR §261.23. There are currently no test methods available.

### Toxicity

Toxic wastes are those containing concentrations of certain substances in excess of regulatory thresholds which are expected to cause injury or illness to human health or the environment.

### Listed Wastes

Listed hazardous wastes are generated by specific industries and processes and are automatically considered hazardous, based solely on the process that generates them and irrespective of whether a test of the waste shows any of the 'characteristics' of hazardous waste. Examples of listed wastes include:

- ➢ many sludges leftover from electroplating processes;
- ➢ certain waste from iron and steel manufacturing;
- ➢ wastes from certain cleaning and/or degreasing processes;

Hazardous wastes are incorporated into lists published by the Environmental Protection Agency. These lists are organized into three categories:

1. *The F-list* (non-specific source wastes): This list identifies wastes from common manufacturing and industrial processes, such as solvents, that have been used in cleaning or degreasing operations. Because the processes producing these wastes can occur in different sectors of industry, the F-listed wastes are known as wastes from non-specific sources.
2. *The K-list* (source-specific wastes): This list includes certain wastes from specific industries, such as petroleum refining or pesticide manufacturing. Certain sludges and wastewaters from treatment and production processes in these industries are examples of source-specific wastes.
3. *Discarded wastes* (*P-List and U-List*): P-List and U-List wastes are actually sublists of the same major list applying to discarded wastes. These wastes apply to commercial chemical products that are considered hazardous when discarded and are regulated under the following U.S. Federal Regulation: 40 C.F.R. 261.33(e) and 261.33(f). P-List wastes are wastes that are considered 'acutely hazardous' when discarded and are subject to more stringent regulation. Nitric oxide is an example of a P-list waste and carries the number P076. U-Listed wastes are

considered 'hazardous' when discarded and are regulated in a somewhat less stringent manner than P-Listed wastes. [Acetone] is an example of a U-Listed waste and carries the number U002.

## Hazardous Waste Mapping Systems

There are several tools for mapping hazardous wastes to particular locations. These tools also allow the user to view additional information.

- TOXMAP is a Geographic Information System (GIS) from the Division of Specialized Information Services of the United States National Library of Medicine (NLM) that uses maps of the United States to help users visually explore data from the United States Environmental Protection Agency's (EPA) Toxics Release Inventory and Superfund Basic Research Programmes. TOXMAP is a resource funded by the US Federal Government. TOXMAP's chemical and environmental health information is taken from NLM's Toxicology Data Network (TOXNET) and PubMed, and from other authoritative sources.
- The US Environmental Protection Agency (EPA) 'Where You Live' allows users to select a region from a map to find information about Superfund sites in that region.

## Universal Wastes

Universal wastes are hazardous wastes that (in the U.S.):

- generally pose a lower threat relative to other hazardous wastes
- are ubiquitous and produced in very large quantities by a large number of generators.

Some of the most common 'universal wastes' are: fluorescent light bulbs, some specialty batteries (e.g. lithium or lead containing batteries), cathode ray tubes, and mercury-containing devices.

Also, in worldwide, The United Nations Environmental Programme (UNEP) estimated that more than 400 million tons of hazardous wastes are produced universally each year, mostly by industrialized countries (schmit, 1999). About one per cent of this total is shipped across international boundaries, with the majority of the transfers occurring between countries in the Organization for the Economic Cooperation and Development(OECD) (Krueger, 1999). In a country like the United States, some undefined portion of the total is shipped legally or illegally to underdeveloped countries. Some of the reasons for industrialized countries to ship the hazardous waste to industrializing countries for disposal are the rising cost of disposing hazardous waste in the home country.

Universal wastes are subject to somewhat less stringent regulatory requirements and small quantity generators of universal wastes may be classified as 'conditionally exempt small quantity generators' (CESQGs) which releases them from some of the regulatory requirements for the handling and storage hazardous wastes.

Universal wastes must still be disposed of properly.

**Other Hazardous Wastes**

The U.S. Environmental Protection Agency has other ways of regulating hazardous waste. These 'rules' include:

- *The 'Mixture Rule'*— applies to a mixture of a listed hazardous waste and a solid waste and states that the result of a mixture of these two wastes is regulated as a hazardous waste. Exemptions may apply in some cases.
- *The 'Derived-from Rule'* — 40 CFR Section 261.3(b) applies to a waste that is generated from the treatment, storage or disposal of a hazardous waste (for example, the ash from the incineration of hazardous waste). Wastes 'derived' in this manner may be regulated as hazardous wastes.
- The 'Contained-in Rule' — applies to soil, groundwater, surface water and debris that are contaminated with a listed hazardous waste.

**Exempted Hazardous Wastes**

USEPA regulations automatically exempt certain solid wastes from being regulated as "hazardous wastes". This does not necessarily mean the wastes are not hazardous nor that they are not regulated. An exempted hazardous waste simply means that the waste is not regulated by the primary hazardous waste regulations. Many of these wastes may by regulated by different statutes and/or regulations and/or by different regulatory agencies. For example, many hazardous mining wastes are regulated via mining statutes and regulations. 'Exempted' hazardous wastes include:

- Household hazardous waste (HHW);
- Agricultural wastes which are returned to the ground as fertilizer;
- Mining overburden returned to the mine site;
- Utility wastes from [coal] combustion to produce electricity;
- Oil and natural gas exploration drilling waste;
- Wastes from the extraction of beneficiation, and processing of ores and minerals, including coal;
- Cement kiln wastes;
- Wood treated with arsenic preservatives;

- Certain chromium-containing wastes;
- Recycled hazardous wastes: Some hazardous wastes that are recycled may also be exempted from hazardous waste regulations.

**Household Hazardous Waste**

Household Hazardous Waste (HHW) (also referred to as domestic hazardous waste) is waste that is generated from residential households. HHW only applies to wastes that are the result of the use of materials that are labeled for and sold for 'home use'.

The following list includes categories often applied to HHW. It is important to note that many of these categories overlap and that many household wastes can fall into multiple categories:

- Paints and solvents
- Automotive wastes (used motor oil, antifreeze, etc.)
- Pesticides ( insecticides, herbicides, fungicides, etc.)
- Mercury-containing wastes ( thermometers, switches, fluorescent lighting, etc.)
- Electronics (computers, televisions, cell phones)
- Aerosols/Propane cylinders
- Caustics/Cleaning agents
- Refrigerant-containing appliances
- Some specialty Batteries (e.g. lithium, nickel cadmium, or button cell batteries)
- Ammunition
- Radioactive waste (some home smoke detectors are classified as radioactive waste because they contain very small amounts of a radioactive isotope of americium.

**Disposal of HHW in the United States**

Because of the expense associated with the disposal of HHW, it is still legal for most homeowners in the U.S. to dispose of most types of household hazardous wastes as municipal solid waste (MSW) and these wastes can be put in your trash. Laws vary by state and municipality and they are changing every day. Be sure to check with your local environmental regulatory agency, solid waste authority, or health department to find out how HHW is managed in your area.

Modern landfills are designed to handle normal amounts of HHW and minimize the environmental impacts. However, there are still going to be some impacts and there are many ways that homeowners can keep these wastes out of landfills.

Laws regulating HHW in the U.S. are gradually becoming more strict. As of 2007, radioactive smoke detectors are the only HHW that are managed nationally. While it is still legal in the United States to dispose of smoke detectors in your trash in most places, manufacturers of smoke detectors must accept returned units for disposal as mandated by the Nuclear Regulatory law 10 CFR 32.27. If you send your detector back to a manufacturer then it will be disposed in a nuclear waste facility.

In the U.S., states are regulating various HHW waste disposal in MSW landfills on a state by state basis. Some commonly regulated wastes in some (but not all) states include restrictions on the disposal of:

- Recyclables (especially 'source-separated' recyclables or recyclables that have already been separated from solid waste). In this case this would only apply to household hazardous wastes that have been separated for recycling.
- Lead-acid batteries
- Mercury-containing wastes
- Rechargeable batteries
- Cathode ray tubes (CRTs) from older computer monitors and televisions
- Cell phones and computers
- Refrigerant containing appliances such as a refrigerator, air conditioner or dehumidifier

(*Note:* Yard waste or 'green waste' (particularly 'source-separated' yard waste such as from a city leaf collection programme) is not hazardous but may be a regulated household waste)

Local solid waste authorities and health departments may also have specific bans on wastes that apply to their service area.

Solid Waste Haulers and HHW — One 'catch-22' that residents often encounter is that while it may be legal to dispose of some HHW in their regular trash, the waste hauler that collects the trash can choose not to haul the waste. It is not uncommon for a waste hauler to refuse to pick up municipal solid waste that contains things like paint and fluorescent light bulbs. There is often little recourse for residents in this case. In these cases the resident may have to make their own arrangements to dispose of the waste by taking it directly to a landfill or solid waste transfer station.

## Final Disposal of Hazardous Waste

Hazardous wastes undergo different treatments in order to stabilize and dispose of them.

### Recycling

Many HHWs can be recycled into new products. Examples might include lead-acid batteries or electronic circuit boards where the heavy metals can be recovered and used in new products. Another example is the ash generated by coal-fired power plants; these plants produced two types of these residues: fly and bottom ash. Fly ash particles have a low density, are very fine, and are removed by air pollution control devices. On the other side, bottom ash is a dense, dark, gravely substance that remains on the bottom of combustion chambers. After these types of ashes go though the proper treatment, they could bind to other pollutants and convert them into easier-to- dispose solids, or they could be used as pavement filling. Such treatments reduce the level of threat of harmful chemicals, like fly and bottom ash, while also recycling the safe product and helping the environment.

### Portland Cement

Another commonly used treatment is cement based solidification and stabilization. Cement is used because it can treat a range of hazardous wastes by improving physical characteristics and decreasing the toxicity and transmission of contaminants. The cement produced is categorized into 5 different divisions, depending on its strength and components. This process of converting sludge into cement might include the addition of pH adjustment agents, phosphates, or sulfur reagents to reduce the settling or curing time, increase the compressive strength, or reduce the leach ability of contaminants.

### Neutralization

Some HW can be processed so that the hazardous component of the waste is eliminated: making it a non-hazardous waste. An example of this might include a corrosive acid that is neutralized with a basic substance so that it is no-longer corrosive. Another mean to neutralize some of the waste is pH adjustment. pH is an important factor on the leaching activity of the hazardous waste. By adjusting the pH of some toxic materials, we are reducing the leaching ability of the waste.

### Incineration, Destruction and Waste-to-energy

A HW may be 'destroyed' for example by incinerating it at a high temperature. Flammable wastes can sometimes be burned as energy sources.

For example many cement kilns burn HWs like used oils or solvents. Today incineration treatments not only reduce the amount of hazardous waste. They also generate energy throughout the gases released in the process. It is known that this particular waste treatment releases toxic gases produced by the combustion of byproduct or other materials and this can affect the environment. However, current technology has developed more efficient incinerator units that control these emissions to a point that this treatment is considered a more beneficial option. There are different types of incinerators and they vary depending on the characteristics of the waste. Starved air incineration is another method used to treat hazardous wastes. Just like in common incineration, burning occurs, however controlling the amount of oxygen allowed proves to be significant to reduce the amount of harmful byproducts produced. Starved Air Incineration is an improvement of the traditional incinerators in terms of air pollution. Using this technology it is possible to control the combustion rate of the waste and therefore reduce the air pollutants produce in the process.

## Hazardous Waste Landfill (Sequestering, Isolation, etc.)

A HW may be sequestered in a HW landfill or permanent disposal facility. "In terms of hazardous waste, a landfill is defined as a disposal facility or part of a facility where hazardous waste is placed or on land and which is not a pile, a land treatment facility, a surface impoundment, an underground injection well, a salt dome formation, a salt bed formation, an underground mine, a cave, or a corrective action management unit (40 CFR 260.10)".

## Pyrolysis

Some hazardous waste types may be eliminated using pyrolisis in an ultra high temperature electrical arc, in inert conditions to avoid combustion. This treatment method may be preferable to high temperature incineration in some circumstances such as in the destruction of concentrated organic waste types, including PCBs, pesticides and other persistent organic pollutants.

## Environmental Dumping

Environmental dumping is the practice of transfrontier shipment of waste (household waste, industrial/nuclear waste, etc.) from one country to another. The goal is to take the waste to a country that has less strict environmental laws, or environmental laws that are not strictly enforced. The economic benefit of this practice is cheap disposal or recycling of waste without the economic regulations of the original country.

An example of an attempt at environmental dumping is the story of the decommissioned French aircraft carrier, the FS Clemenceau, which was sold to India to be scrapped. The Indian Supreme Court ruled that it could not enter Indian waters due to the high level of toxic waste and asbestos found on the ship, forcing the French government to take the Clemenceau back.

The shipment of waste between countries has been referred to as 'transfrontier shipment' of waste. Transfrontier waste is shipped within the European Union (EU) and between the European Union and other countries. Most of this waste is traded by Organization for Economic Co-operation and Development (OECD) countries. The waste is typically non-hazardous and includes metals, plastics, and paper products. In 2007, it is estimated that OECD countries exported between 4 and 5 million tons of metal and paper waste. OECD countries also exported near a half of million tons of recovered plastics in 2007. Some of these wastes that are transported can be hazardous waste. These hazardous wastes can cause potential health risks to humans and the environment. According to the Basel Convention, there is at least 8 million tons of hazardous waste imported and exported every year.

The Basel Convention was created in 1989 but started enforcing rules in 1992. The purpose of the Convention is to control the hazardous waste that was imported and exported throughout the EU. The Convention is a great contributor to stopping the shipment of illegal waste. In May 2005, 60 containers were seized that were on their way from the United Kingdom to China. The containers seized by Dutch authorities were supposed to be for paper but actually contained household wastes. Since neither the UK, China, nor Dutch had agreed to the importation of the wastes, the waste was shipped back. The Basel Convention also deals with the popular growing issue of E-waste. The Waste Shipment Regulation confirms what can be shipped to, from, and between EU countries. These regulation rules divide the waste into three separate lists: Green List, Amber List, and Red List.

**Green List**

These items are considered to be non-hazardous and more environmentally friendly. Some of these items may include paper and plastic that can be recycled. These types of shipment don't have to receive prior permission to cross international waters and be shipped to parts of the European Union.

**Amber List**

Materials are considered to be mixed on this list containing both non-hazardous and hazardous parts. These materials can contain metal bearing wastes, organic and inorganic wastes, and/or organic or inorganic constituents. A company or country shipping these items would have to have prior consent before exporting the materials. As of 2007, consent for the shipment of waste is received by Dublin City Council.

### Red List

This includes reasonably hazardous materials. These materials contain principally organic or inorganic constituents which include polychlorinated biphenyl (PCBs).

### Ocean Dumping

Shipment of waste from country to country can also involve dumping waste into the ocean. Ocean dumping has been a problem since the 19th century. In the past, it was legal to dump industrial waste into the ocean until the Ocean Dumping Act was passed in 1972. During the years of 1970 and 1980 alone, it was estimated that 25 million tons of waste including scrap metal, chemicals, and acids were dumped into the ocean. Ocean dumping can lead to eutrophication which depletes the oxygen from the water killing marine life. Ocean dumping is placed into three lists: Gray List, Black List, and White List.

### Gray List

Water is highly contaminated and toxic on the gray list. Some of the contaminants include arsenic, lead, acids, nickel, chromium, scrap metals, and radioactive materials.

### Black List

The materials listed on the black list include mercury, cadmium, plastic, oil products, radioactive waste, and anything that is solely made for biological and chemical warfare. The materials on this list are highly potent and hazardous.

### White List

The white list contains every other material not already mentioned in the above lists. This list is used to make sure nothing will be dumped in the ocean and disturb or damage the coral reef ecosystems.

The most recent example of hazardous waste being dumped into the ocean occurred along the Ivory Coast in Africa in 2006. Hundreds of tons of toxic waste were dumped into the ocean from a ship by the name of Probo Koala. The ship was chartered by an international oil trader in the Netherlands. The incident was responsible for killing at least 16 people and forced thousands to flee from their homes. The waste caused damages to the health of almost 100,000 people in the nearby areas . The oil trader by the name of Trafigura paid 200 million dollars to help with cleanup of the

Ivory Coast . The owner of the local company, Tommy that was responsible for disposing the chemicals in various places was charged with 20 years jail time. The man who declared that Tommy was an accomplice in the situation charged with 5 years jail time for being involved.

After this incident some new regulations on the shipment of hazardous and non-hazardous materials were implemented in 2006. According to the new regulations, the EU will no longer be able to export their hazardous wastes to developing countries that do not have the capabilities to deal with the waste in an environmentally friendly way. E-waste, such as computers, cannot be shipped to countries that are not in the EU or the European Free Trade Association. States and countries that are members of the EU or EFTA must conduct inspections periodically to make sure that all regulations are being followed and physically check containers to verify that the containers hold only what is authorized. Finally, if the country that is taking the waste is unable to accept or dispose of the waste then the sender must pay to take their waste back. The purpose of these changes is to make sure that incidents like the Ivory Coast doesn't happen again.

## Ship Dismantling

Ship dismantling is another form of transfrontier waste that is shipped from country to country. Many ships are broken down into parts that can be recycled. Many parts of the ships are hazardous and can potentially pollute the areas that they are broken down in. The ship parts can contain asbestos, PCBs, and oil sludge. All of these components can be a potential health risk and harm the environment. Most ship scrapping industries are in developing countries where the laws (environmentally as well as occupationally) are not as strict as in developed countries. International Maritime Organization states that India is the leader in ship dismantling, followed by China, Bangladesh, and Pakistan. The aircraft carrier Clemenceau was denied access to Indian waters because it contained asbestos. The French aircraft was carried from France to Britain to be recycled on February 8, 2009 despite the abundance of asbestos.

The EU Commission proposed improvements to be made to improve the ship dismantling as well as the ocean dumping process on November 19, 2008. These improvements were deemed necessary after the dumping incident on Ivory Coast in 2006. One major improvement would be to provide facilities designed to break down ship parts safely for recycling mostly in regions of South Asia. The Ship Recycling Convention will not decide on the improvements until May 2009. If the new improvements for making ship dismantling safer for the areas and workers who dismantle the parts are agreed upon at the convention, it would take years before it became international law.

CHAPTER 9

# Electronic Waste

Electronic waste, e-waste, e-scrap, or Waste Electrical and Electronic Equipment (WEEE) describes loosely discarded, surplus, obsolete, or broken electrical or electronic devices. Environmental groups claim that the informal processing of electronic waste in developing countries causes serious health and pollution problems. Some electronic scrap components, such as CRTs, contain contaminants such as lead, cadmium, beryllium, mercury, and brominated flame retardants. Activists claim that even in developed countries recycling and disposal of e-waste may involve significant risk to workers and communities and great care must be taken to avoid unsafe exposure in recycling operations and leaching of material such as heavy metals from landfills and incinerator ashes. Scrap industry and USA EPA officials agree that materials should be managed with caution, but that environmental dangers of unused electronics have been exaggerated by groups which benefit from increased regulation.

'Electronic waste' may be defined as all secondary computers, entertainment device electronics, mobile phones, and other items such as television sets and refrigerators, whether sold, donated, or discarded by their original owners. This definition includes used electronics which are destined for reuse, resale, salvage, recycling, or disposal. Others define the re-usables (working and repairable electronics) and secondary scrap (copper, steel, plastic, etc.) to be 'commodities', and reserve the term 'waste' for residue or material which was represented as working or repairable but which is dumped or disposed or discarded by the buyer rather than recycled, including residue from reuse and recycling operations. Because loads of surplus electronics are frequently commingled (good, recyclable, and non-recyclable), several public policy advocates apply the term 'e-waste' broadly to all surplus electronics. The United States Environmental Protection Agency (EPA) includes discarded CRT monitors in its category of 'hazardous household waste'. but considers CRTs set aside for testing to be commodities if they are not discarded, speculatively accumulated, or left unprotected from weather and other damage.

Debate continues over the distinction between 'commodity' and 'waste' electronics definitions. Some exporters may deliberately leave difficult-to-spot obsolete or non-working equipment mixed in loads of working equipment (through ignorance, or to avoid more costly treatment processes). Protectionists may broaden the definition of 'waste' electronics. The high value of the computer recycling subset of electronic waste (working and reusable laptops, computers, and components like RAM) can help pay the cost of transportation for a large number of worthless 'electronic commodities'.

Rapid changes in technology, low initial cost, and planned obsolescence have resulted in a fast-growing surplus of electronic waste around the globe. Dave Kruch, CEO of Cash For Laptops, regards electronic waste as a "rapidly expanding" issue. Technical solutions are available, but in most cases a legal framework, a collection system, logistics, and other services need to be implemented before a technical solution can be applied. An estimated 50 million tons of E-waste is produced each year. The USA discards 30 million computers each year and 100 million phones are disposed of in Europe each year. The Environmental Protection Agency estimates that only 15-20 per cent of e-waste is recycled, the rest of these electronics go directly into landfills and incinerators.

In the United States, an estimated 70 per cent of heavy metals in landfills comes from discarded electronics.

Increased regulation of electronic waste and concern over the environmental harm which can result from toxic electronic waste has raised disposal costs. The regulation creates an economic disincentive to remove residues prior to export. Critics of trade in used electronics maintain that it is too easy for brokers calling themselves recyclers to export unscreened electronic waste to developing countries, such as China, India and parts of Africa, thus avoiding the expense of removing items like bad cathode ray tubes (the processing of which is expensive and difficult). The developing countries are becoming big dump yards of e-waste due to their weak laws. Proponents of international trade point to the success of fair trade programs in other industries, where cooperation has led creation of sustainable jobs, and can bring affordable technology in countries where repair and reuse rates are higher.

Defenders of the trade in used electronics say that extraction of metals from virgin mining has also been shifted to developing countries. Hard-rock mining of copper, silver, gold and other materials extracted from electronics is considered far more environmentally damaging than the recycling of those materials. They also state that repair and reuse of computers and televisions has become a "lost art" in wealthier nations, and that refurbishing has traditionally been a path to development. South Korea, Taiwan, and southern China all excelled in finding 'retained value' in used goods, and in some cases have set up billion-dollar industries in refurbishing used ink cartridges, single-use cameras, and working CRTs. Refurbishing has traditionally been a threat to established manufacturing, and simple protectionism explains some criticism of the trade. Works like 'The Waste Makers' by Vance Packard explain some of the criticism of exports of working product, for example the ban on import of tested working Pentium 4 laptops to China, or the bans on export of used surplus working electronics by Japan.

Opponents of surplus electronics exports argue that lower environmental and labour standards, cheap labour, and the relatively high value of recovered raw materials leads to a transfer of pollution-generating activities, such as burning of copper wire. In China, Malaysia, India, Kenya, and various African countries, electronic waste is being sent to these countries for processing, sometimes illegally. Many surplus laptops are routed to developing nations as 'dumping grounds for e-waste'. Because the United States has not ratified the Basel Convention or its Ban Amendment, and has no domestic laws forbidding the export of toxic waste, the Basel Action Network estimates that about 80 per cent of the electronic waste directed to recycling in the U.S. does not get recycled there at all, but is put on container ships and sent to countries such as China. This figure is disputed as an exaggeration by the EPA, the Institute for Scrap Recycling Industries, and the World Reuse, Repair and Recycling Association. Independent research by Arizona State University showed that 87-88 per cent of imported used computers had a higher value than the best value of the constituent materials they contained, and that "the official trade in end-of-life computers is thus drive by reuse as opposed to recycling".

Guiyu in the Shantou region of China, Delhi and Bangalore in India as well as the Agbogbloshie site near Accra, Ghana have electronic waste processing areas. Uncontrolled burning, disassembly, and disposal can cause a variety of environmental problems such as groundwater contamination, atmospheric pollution, or even water pollution either by immediate discharge or due to surface runoff (especially near coastal areas), as well as health problems including occupational safety and health effects among those directly involved, due to the methods of processing the waste. Thousands of men, women, and children are employed in highly polluting, primitive recycling technologies, extracting the metals, toners, and plastics from computers and other electronic waste. Recent studies show that 7 out of 10 children in this region have too much lead in their blood.

Proponents of the trade say growth of internet access is a stronger correlation to trade than poverty. Haiti is poor and closer to the port of New York than southeast Asia, but far more electronic waste is exported from New York to Asia than to Haiti. Thousands of men, women, and children are employed in reuse, refurbishing, repair, and remanufacturing, sustainable industries in decline in developed countries. It is held that denying developing nations access to used electronics denies them affordable products and internet access.

Opponents of the trade argue that developing countries utilize methods that are more harmful and more wasteful. An expedient and prevalent method is simply to toss equipment onto an open fire, in order to melt plastics and to burn away unvaluable metals. This releases carcinogens and neurotoxins into the air, contributing to an acrid, lingering smog. These noxious fumes include dioxins and furans. Bonfire refuse can be disposed of quickly into drainage ditches or waterways feeding the ocean or local water supplies.

In June 2008, a container of electronic waste, destined from the Port of Oakland in the U.S. to Sanshui District in mainland China, was intercepted in Hong Kong by Greenpeace. Concern over exports of electronic waste were raised in press reports in India, Ghana, Ivory Coast, and Nigeria.

Today the electronic waste recycling business is in all areas of the developed world a large and rapidly consolidating business. Electronic waste processing systems have matured in recent years, following increased regulatory, public, and commercial scrutiny, and a commensurate increase in entrepreneurial interest. Part of this evolution has involved greater diversion of electronic waste from energy-intensive downcycling processes (e.g., conventional recycling), where equipment is reverted to a raw material form. This diversion is achieved through reuse and refurbishing. The environmental and social benefits of reuse include diminished demand for new products and virgin raw materials (with their own environmental issues); larger quantities of pure water and electricity for associated manufacturing; less packaging per unit; availability of technology to wider swaths of society due to greater affordability of products; and diminished use of landfills.

Audiovisual components, televisions, VCRs, stereo equipment, mobile phones, other handheld devices, and computer components contain valuable elements and substances suitable for reclamation, including lead, copper, and gold.

One of the major challenges is recycling the printed circuit boards from the electronic wastes. The circuit boards contain such precious metals as gold, silver, platinum, etc. and such base metals as copper, iron, aluminum, etc. Conventional method employed is mechanical shredding and separation but the recycling efficiency is low. Alternative methods such as cryogenic decomposition have been studied for printed circuit board recycling, and some other methods are still under investigation.

## Consumer Awareness Efforts

Address the mess.com is a Comedy Central pro-social campaign that seeks to increase awareness of the dangers of electronic waste and to encourage recycling. Partners in the effort include Earth911.org, ECOInternational.com, and the U.S. Environmental Protection Agency?. Many Comedy Central viewers are early adopters of new electronics, and produce a commensurate amount of waste that can be directed towards recycling efforts. The station is also taking steps to reduce its own environmental impact, in partnership with NativeEnergy.com, a company that specializes in renewable energy and carbon offsets.

The Electronics TakeBack Coalition is a campaign aimed at protecting human health and limiting environmental effects where electronics are being produced, used, and discarded. The ETBC aims to place responsibility for disposal of technology products on electronic manufacturers and brand owners, primarily through community promotions and legal enforcement initiatives. It provides recommendations for consumer recycling and a list of recyclers judged environmentally responsible.

The grassroots Silicon Valley Toxics Coalition (svtc.org) focuses on promoting human health and addresses environmental justice problems resulting from toxins in technologies.

Basel Action Network (BAN.org) is uniquely focused on addressing global environmental injustices and economic inefficiency of global 'toxic trade'. It works for human rights and the environment by preventing disproportionate dumping on a large scale. It promotes sustainable solutions and attempts to ban waste trade.

Texas Campaign for the Environment (texasenvironment.org) works to build grassroots support for e-waste recycling and uses community organizing to pressure electronics manufacturers and elected officials to enact producer takeback recycling policies and commit to responsible recycling programmes.

The World Reuse, Repair, and Recycling Association is an organization dedicated to improving the quality of exported electronics, encouraging better recycling standards in importing countries, and improving practices through 'Fair Trade' principles.

Take Back My TV is a project of The Electronics TakeBack Coalition and grades television manufacturers to find out which are responsible and which are not.

In developed countries, electronic waste processing usually first involves dismantling the equipment into various parts (metal frames, power supplies, circuit boards, plastics), often by hand. The advantages of this process are the human's ability to recognize and save working and repairable parts, including chips, transistors, RAM, etc. The disadvantage is that the labour is often cheapest in countries with the lowest health and safety standards.

In an alternative bulk system, a hopper conveys material for shredding into a sophisticated mechanical separator, with screening and granulating machines to separate constituent metal and plastic fractions, which are sold to smelters or plastics recyclers. Such recycling machinery is enclosed and employs a dust collection system. Most of the emissions are caught by scrubbers and screens. Magnets, eddy currents, and trommel screens are employed to separate glass, plastic, and ferrous and nonferrous metals, which can then be further separated at a smelter. Leaded glass from CRTs is reused in car batteries, ammunition, and lead wheel weights, or sold to foundries as a fluxing agent in processing raw lead ore. Copper, gold, palladium, silver, and tin are valuable metals sold to smelters for recycling. Hazardous smoke and gases are captured, contained, and treated to mitigate environmental threat. These methods allow for safe reclamation of all valuable computer construction materials. Hewlett-Packard product recycling solutions manager Renee St. Denis describes its process as: "We move them through giant shredders about 30 feet tall and it shreds everything into pieces about the size of a quarter. Once your disk drive is shredded into pieces about this big, it's hard to get the data off".

An ideal electronic waste recycling plant combines dismantling for component recovery with increased cost-effective processing of bulk electronic waste.

Reuse is an option to recycling because it extends the lifespan of a device. Devices still need eventual recycling, but by allowing others to purchase used electronics, recycling can be postponed and value gained from device use.

### Benefits of Recycling

Recycling raw materials from end-of-life electronics is the most effective solution to the growing e-waste problem. Most electronic devices contain a variety of materials, including metals that can be recovered for future uses. By dismantling and providing reuse possibilities, intact natural resources are conserved and air and water pollution caused by hazardous disposal is avoided. Additionally, recycling reduces the amount of greenhouse gas emissions caused by the manufacturing of new products. It simply makes good sense and is efficient to recycle and to do our part to keep the environment green.

Some computer components can be reused in assembling new computer products, while others are reduced to metals that can be reused in applications as varied as construction, flatware, and jewellery.

Substances found in large quantities include epoxy resins, fibreglass, PCBs, PVC (polyvinyl chlorides), thermosetting plastics, lead, tin, copper, silicon, beryllium, carbon, iron and aluminium.

Elements found in small amounts include cadmium, mercury, and thallium.

Elements found in trace amounts include americium, antimony, arsenic, barium, bismuth, boron, cobalt, europium, gallium, germanium, gold, indium, lithium, manganese, nickel, niobium, palladium, platinum, rhodium, ruthenium, selenium, silver, tantalum, terbium, thorium, titanium, vanadium, and yttrium.

Almost all electronics contain lead and tin (as solder) and copper (as wire and printed circuit board tracks), though the use of lead-free solder is now spreading rapidly. The following are ordinary applications:

**Hazardous**

- Americium: smoke alarms (radioactive source).
- Mercury: fluorescent tubes (numerous applications), tilt switches (pinball games, mechanical doorbells, thermostats). There are no liquid mercury switches in ordinary computers, and the elimination of mercury batteries in many new-model computers is taking place.
- Sulphur: lead-acid batteries.
- PBBs: Predecessor of PCBs. Also used as flame retardant. Banned from 1973-1977 on.
- PCBs: prior to ban, almost all 1930s-1970s equipment, including capacitors, transformers, wiring insulation, paints, inks, and flexible sealants. Banned during the 1980s.
- Cadmium: light-sensitive resistors, corrosion-resistant alloys for marine and aviation environments, nickel-cadmium batteries.
- Lead: solder, CRT monitor glass, lead-acid batteries, some formulations of PVC. A typical 15-inch cathode ray tube may contain 1.5 pounds of lead, but other CRTs have been estimated as having up to 8 pounds of lead.

- Beryllium oxide: filler in some thermal interface materials such as thermal grease used on heatsinks for CPUs and power transistors, magnetrons, X-ray-transparent ceramic windows, heat transfer fins in vacuum tubes, and gas lasers.
- Polyvinyl chloride Third most widely produced plastic, contains additional chemicals to change the chemical consistency of the product. Some of these additional chemicals called additives can leach out of vinyl products. Plasticizers that must be added to make PVC flexible have been additives of particular concern. Burning PVC in connection with humidity in the air creates Hydrogen Chloride (HCl), an acid.

**Generally Non-hazardous**

- Tin: solder, coatings on component leads.
- Copper: copper wire, printed circuit board tracks, component leads.
- Aluminium: nearly all electronic goods using more than a few watts of power (heatsinks), electrolytic capacitors.
- Iron: steel chassis, cases, and fixings.
- Germanium: 1950s-1960s transistorized electronics (bipolar junction transistors).
- Silicon: glass, transistors, ICs, printed circuit boards.
- Nickel: nickel-cadmium batteries.
- Lithium: lithium-ion batteries.
- Zinc: plating for steel parts.
- Gold: connector plating, primarily in computer equipment.

# Chapter 10

# Waste Management

Waste management is the collection, transport, processing, recycling or disposal, and monitoring of waste materials. The term usually relates to materials produced by human activity, and is generally undertaken to reduce their effect on health, the environment or aesthetics. Waste management is also carried out to recover resources from it. Waste management can involve solid, liquid, gaseous or radioactive substances, with different methods and fields of expertise for each.

Waste management practices differ for developed and developing nations, for urban and rural areas, and for residential and industrial producers. Management for non-hazardous residential and institutional waste in metropolitan areas is usually the responsibility of local government authorities, while management for non-hazardous commercial and industrial waste is usually the responsibility of the generator.

**Integrated Waste Management**

Integrated waste management using LCA (life cycle analysis) attempts to offer the most benign options for waste management. For mixed MSW (Municipal Solid Waste) a number of broad studies have indicated that waste administration, then source separation and collection followed by reuse and recycling of the non-organic fraction and energy and compost/ fertilizer production of the organic waste fraction via anaerobic digestion to be the favoured path. Non-metallic waste resources are not destroyed as with incineration, and can be reused/recycled in a future resource depleted society.

## Plasma Gasification

Plasma is a highly ionized or electrically charged gas. An example in nature is lightning, capable of producing temperatures exceeding 12,600° F (6,980° C). A gasifier vessel utilizes proprietary plasma torches operating at +10,000° F (5,540° C) (the surface temperature of the Sun) in order to create a gasification zone of up to 3,000° F (1,650° C) to convert solid or liquid wastes into a syngas. When municipal solid waste is subjected to this intense heat within the vessel, the waste's molecular bonds break down into elemental components. The process results in elemental destruction of waste and hazardous materials.

According to the U.S. Environmental Protection Agency, the U.S. generated 250 million tons of waste in 2008 alone, and this number continues to rise. About 54 per cent of this trash (135,000,000 short tons (122,000,000 t)) ends up in landfills and is consuming land at a rate of nearly 3,500 acres (1,400 ha) per year. In fact, landfilling is currently the number one method of waste disposal in the US. Some states no longer have capacity at permitted landfills and export their waste to other states. Plasma gasification offers states new opportunities for waste disposal, and more importantly for renewable power generation in an environmentally sustainable manner.

Disposing of waste in a landfill involves burying the waste, and this remains a common practice in most countries. Landfills were often established in abandoned or unused quarries, mining voids or borrow pits. A properly designed and well-managed landfill can be a hygienic and relatively inexpensive method of disposing of waste materials. Older, poorly designed or poorly managed landfills can create a number of adverse environmental impacts such as wind-blown litter, attraction of vermin, and generation of liquid leachate. Another common byproduct of landfills is gas (mostly composed of methane and carbon dioxide), which is produced as organic waste breaks down anaerobically. This gas can create odour problems, kill surface vegetation, and is a greenhouse gas.

Design characteristics of a modern landfill include methods to contain leachate such as clay or plastic lining material. Deposited waste is normally compacted to increase its density and stability, and covered to prevent attracting vermin (such as mice or rats). Many landfills also have landfill gas extraction systems installed to extract the landfill gas. Gas is pumped out of the landfill using perforated pipes and flared off or burnt in a gas engine to generate electricity.

Incineration is a disposal method in which solid organic wastes are subjected to combustion so as to convert them into residue and gaseous products.This method is useful for disposal of residue of both solid waste management and solid residue from waste water management.This process reduces the volumes of solid waste to 20 to 30 per cent of the original volume. Incineration and other high temperature waste treatment systems are sometimes described as 'thermal treatment'. Incinerators convert waste materials into heat, gas, steam and ash.

Incineration is carried out both on a small scale by individuals and on a large scale by industry. It is used to dispose of solid, liquid and gaseous waste. It is recognized as a practical method of disposing of certain hazardous waste materials (such as biological medical waste). Incineration is a controversial method of waste disposal, due to issues such as emission of gaseous pollutants.

Incineration is common in countries such as Japan where land is more scarce, as these facilities generally do not require as much area as landfills. Waste-to-energy (WtE) or energy-from-waste (EfW) are broad terms for facilities that burn waste in a furnace or boiler to generate heat, steam and/or electricity. Combustion in an incinerator is not always perfect and there have been concerns about micro-pollutants in gaseous emissions from incinerator stacks. Particular concern has focussed on some very persistent organics such as dioxins, furans, PAHs, . . . which may be created within the incinerator and afterwards in the incinerator plume which may have serious environmental consequences in the area immediately around the incinerator. On the other hand this method or the more benign anaerobic digestion produces heat that can be used as energy.

The popular meaning of 'recycling' in most developed countries refers to the widespread collection and reuse of everyday waste materials such as empty beverage containers. These are collected and sorted into common types so that the raw materials from which the items are made can be reprocessed into new products. Material for recycling may be collected separately from general waste using dedicated bins and collection vehicles, or sorted directly from mixed waste streams.

The most common consumer products recycled include aluminum beverage cans, steel food and aerosol cans, HDPE and PET bottles, glass bottles and jars, paperboard cartons, newspapers, magazines, and corrugated fibreboard boxes.

PVC, LDPE, PP, and PS (see resin identification code) are also recyclable, although these are not commonly collected. These items are usually composed of a single type of material, making them relatively easy to recycle into new products. The recycling of complex products (such as computers and electronic equipment) is more difficult, due to the additional dismantling and separation required.

**Sustainability**

The management of waste is a key component in a business' ability to maintaining ISO14001 accreditations. Companies are encouraged to improve their environmental efficiencies each year. One way to do this is by improving a company's waste management with a new recycling service.

Waste materials that are organic in nature, such as plant material, food scraps, and paper products, can be recycled using biological composting and digestion processes to decompose the organic matter. The resulting organic material is then recycled as mulch or compost for agricultural or landscaping purposes. In addition, waste gas from the process (such as methane) can be captured and used for generating electricity and heat (CHP/cogeneration) maximising efficiencies. The intention of biological processing in waste management is to control and accelerate the natural process of decomposition of organic matter.

There are a large variety of composting and digestion methods and technologies varying in complexity from simple home compost heaps, to small town scale batch digesters, industrial-scale enclosed-vessel digestion of mixed domestic waste. Methods of biological decomposition are differentiated as being aerobic or anaerobic methods, though hybrids of the two methods also exist.

Anaerobic digestion of the organic fraction of MSW Municipal Solid Waste has been found to be in a number of LCA analysis studies to be more environmentally effective, than landfill, incineration or pyrolisis. The resulting biogas (methane) though must be used for cogeneration (electricity and heat preferably on or close to the site of production) and can be used with a little upgrading in gas combustion engines or turbines. With further upgrading to synthetic natural gas it can be injected into the natural gas network or further refined to hydrogen for use in stationary cogeneration fuel cells. Its use in fuel cells eliminates the pollution from products of combustion (SOx, NOx, pariculates, dioxin, furans, PAHs. . .).

An example of waste management through composting is the Green Bin Programme in Toronto, Canada, where household organic waste (such as kitchen scraps and plant cuttings) are collected in a dedicated container and then composted.

The energy content of waste products can be harnessed directly by using them as a direct combustion fuel, or indirectly by processing them into another type of fuel. Recycling through thermal treatment ranges from using waste as a fuel source for cooking or heating, to anaerobic digestion and the use of the gas fuel, to fuel for boilers to generate steam and electricity in a turbine. Pyrolysis and gasification are two related forms of thermal treatment where waste materials are heated to high temperatures with limited oxygen availability. The process usually occurs in a sealed vessel under high pressure. Pyrolysis of solid waste converts the material into solid, liquid and gas products. The liquid and gas can be burnt to produce energy or refined into other chenmical products (chemical refinery). The solid residue (char) can be further refined into products such as activated carbon. Gasification and advanced Plasma arc gasification are used to convert organic materials directly into a synthetic gas (syngas) composed of carbon monoxide and hydrogen. The gas is then burnt to produce electricity and steam. An alternative to pyrolisis is high temperature and pressure supercritical water decomposition (hydrothermal monophasic oxidation).

## Avoidance and Reduction Methods

An important method of waste management is the prevention of waste material being created, also known as waste reduction. Methods of avoidance include reuse of second-hand products, repairing broken items instead of buying new, designing products to be refillable or reusable (such as cotton instead of plastic shopping bags), encouraging consumers to avoid using disposable products (such as disposable cutlery), removing any food/liquid remains from cans, packaging, . . . and designing products that use less material to achieve the same purpose.

## Waste Handling and Transport

Waste collection methods vary widely among different countries and regions. Domestic waste collection services are often provided by local government authorities, or by private companies in the industry. Some areas, especially those in less developed countries, do not have a formal waste-collection system. Examples of waste handling systems include:

- In Australia, curbside collection is the method of disposal of waste. Every urban domestic household is provided with three bins: one for recyclables, another for general waste and another for garden materials — this bin is provided by the municipality if requested. Also, many households have compost bins; but this is not provided by the municipality. To encourage recycling, municipalities provide large recycle bins, which are larger than general waste bins. Municipal, commercial and industrial, construction and demolition waste is dumped at landfills and some is recycled. Household waste is segregated: recyclables sorted and made into new products, and general waste is dumped in landfill areas. According to the ABS, the recycling rate is high and is 'increasing, with 99 per cent of households reporting that they had recycled or reused some of their waste within the past year (2003 survey), up from 85 per cent in 1992'. This suggests that

Australians are in favour of reduced or no landfilling and the recycling of waste. Of the total waste produced in 2002-03, '30 per cent of municipal waste, 45 per cent of commercial and industrial waste and 57 per cent of construction and demolition waste' was recycled. Energy is produced from waste as well: some landfill gas is captured for fuel or electricity generation. Households and industries are not charged for the volume of waste they produce.

- In Europe and a few other places around the world, a few communities use a proprietary collection system known as Envac, which conveys refuse via underground conduits using a vacuum system. Other vacuum-based solutions include the MetroTaifun single-line and ring-line systems.
- In Canadian urban centres curbside collection is the most common method of disposal, whereby the city collects waste and/or recyclables and/or organics on a scheduled basis. In rural areas people often dispose of their waste by hauling it to a transfer station. Waste collected is then transported to a regional landfill.
- In Taipei, the city government charges its households and industries for the volume of rubbish they produce. Waste will only be collected by the city council if waste is disposed in government issued rubbish bags. This policy has successfully reduced the amount of waste the city produces and increased the recycling rate.
- In Israel, the Arrow Ecology company has developed the ArrowBio system, which takes trash directly from collection trucks and separates organic and inorganic materials through gravitational settling, screening, and hydro-mechanical shredding. The system is capable of sorting huge volumes of solid waste, salvaging recyclables, and turning the rest into biogas and rich agricultural compost. The system is used in California, Australia, Greece, Mexico, the United Kingdom and in Israel. For example, an ArrowBio plant that has been operational at the Hiriya landfill site since December 2003 serves the Tel Aviv area, and processes up to 150 tons of garbage a day.

**Technologies**

Traditionally the waste management industry has been slow to adopt new technologies such as RFID (Radio Frequency Identification) tags, GPS and integrated software packages which enable better quality data to be collected without the use of estimation or manual data entry.

Technologies like RFID tags are now being used to collect data on presentation rates for curb-side pick-ups which is useful when examining the usage of recycling bins or similar.

Benefits of GPS tracking is particularly evident when considering the efficiency of ad hoc pick-ups (like skip bins or dumpsters) where the collection is done on a consumer request basis.

Integrated software packages are useful in aggregating this data for use in optimisation of operations for waste collection operations.

Rear vision cameras are commonly used for OH&S reasons and video recording devices are becoming more widely used, particularly concerning residential services and contaminations of the waste stream.

## Waste Management Concepts

There are a number of concepts about waste management which vary in their usage between countries or regions. Some of the most general, widely used concepts include:

- *Waste hierarchy* — The waste hierarchy refers to the '3 Rs' reduce, reuse and recycle, which classify waste management strategies according to their desirability in terms of waste minimization. The waste hierarchy remains the cornerstone of most waste minimization strategies. The aim of the waste hierarchy is to extract the maximum practical benefits from products and to generate the minimum amount of waste.
- *Extended producer responsibility* — Extended Producer Responsibility (EPR) is a strategy designed to promote the integration of all costs associated with products throughout their life cycle (including end-of-life disposal costs) into the market price of the product. Extended producer responsibility is meant to impose accountability over the entire lifecycle of products and packaging introduced to the market. This means that firms which manufacture, import and/or sell products are required to be responsible for the products after their useful life as well as during manufacture.
- *Polluter pays principle* — The Polluter Pays Principle is a principle where the polluting party pays for the impact caused to the environment. With respect to waste management, this generally refers to the requirement for a waste generator to pay for appropriate disposal of the waste.

Education and awareness in the area of waste and waste management is increasingly important from a global perspective of resource management. The Talloires Declaration is a declaration for sustainability concerned about the unprecedented scale and speed of environmental pollution and degradation, and the depletion of natural resources. Local, regional, and global air pollution; accumulation and distribution of toxic wastes; destruction and depletion of forests, soil, and water; depletion of the ozone layer and emission of 'green house' gases threaten the survival of humans and thousands of other living species, the integrity of the earth and its biodiversity, the security of nations, and the heritage of future generations. Several universities have implemented the Talloires Declaration by establishing environmental management and waste management programmes, e.g. the waste management university project. University and vocational education are promoted by various organizations, e.g. WAMITAB and Chartered Institution of Wastes Management. Many supermarkets encourage customers to use their reverse vending machines to deposit used purchased containers and receive a refund from the recycling fees. Brands that manufacture such machines include Tomra and Envipco.

# Chapter 11
# Recycling

Recycling involves processing used materials (waste) into new products to prevent waste of potentially useful materials, reduce the consumption of fresh raw materials, reduce energy usage, reduce air pollution (from incineration) and water pollution (from landfilling) by reducing the need for 'conventional' waste disposal, and lower greenhouse gas emissions as compared to virgin production. Recycling is a key component of modern waste reduction and is the third component of the 'Reduce, Reuse, Recycle' waste hierarchy.

Recyclable materials include many kinds of glass, paper, metal, plastic, textiles, and electronics. Although similar in effect, the composting or other reuse of biodegradable waste — such as food or garden waste — is not typically considered recycling. Materials to be recycled are either brought to a collection center or picked up from the curbside, then sorted, cleaned, and reprocessed into new materials bound for manufacturing.

In a strict sense, recycling of a material would produce a fresh supply of the same material—for example, used office paper would be converted into new office paper, or used foamed polystyrene into new polystyrene. However, this is often difficult or too expensive (compared with producing the same product from raw materials or other sources), so 'recycling' of many products or materials involves their reuse in producing different materials (e.g., paperboard) instead. Another form of recycling is the salvage of certain materials from complex products, either due to their intrinsic value (e.g., lead from car batteries, or gold from computer components), or due to their hazardous nature (e.g., removal and reuse of mercury from various items). Critics dispute the net economic and environmental benefits of recycling over its costs, and suggest that proponents of recycling often make matters worse and suffer from confirmation bias. Specifically, critics argue that the costs and energy used in collection and transportation detract from (and outweigh) the costs and energy saved in the production process; also that the jobs produced by the recycling industry can be a poor trade for the jobs lost in logging, mining, and other industries associated with virgin production; and that materials such as paper pulp can only be recycled a few times before material degradation prevents further recycling. Proponents of recycling dispute each of these claims, and the validity of arguments from both sides has led to enduring controversy.

Recycling has been a common practice for most of human history, with recorded advocates as far back as Plato in 400 BC. During periods when utilities were scarce, archaeological studies of ancient waste dumps show less household waste (such as ash, broken tools and pottery)—implying more waste was being recycled in the absence of new material.

In pre-industrial times, there is evidence of scrap bronze and other metals being collected in Europe and melted down for perpetual reuse. In Britain dust and ash from wood and coal fires was collected by 'dustmen' and downcycled as a base material used in brick making. The main driver for these types of recycling was the economic advantage of obtaining recycled feedstock instead of acquiring virgin material, as well as a lack of public waste removal in ever more densely populated areas. In 1813, Benjamin Law developed the process of turning rags into 'shoddy' and 'mungo' wool in Batley, Yorkshire. This material combined recycled fibres with virgin wool. The West Yorkshire shoddy industry in towns such as Batley and Dewsbury, lasted from the early 19th century to at least 1914.

Industrialization spurred demand for affordable materials; aside from rags, ferrous scrap metals were coveted as they were cheaper to acquire than was virgin ore. Railroads both purchased and sold scrap metal in the 19th century, and the growing steel and automobile industries purchased scrap in the early 20th century. Many secondary goods were collected, processed, and sold by peddlers who combed dumps, city streets, and went door to door looking for discarded machinery, pots, pans, and other sources of metal. By World War I, thousands of such peddlers roamed the streets of American cities, taking advantage of market forces to recycle post-consumer materials back into industrial production.

## Wartime Recycling

Resource shortages caused by the world wars, and other such world-changing occurrences greatly encouraged recycling. Massive government promotion campaigns were carried out in World War II in every country involved in the war, urging citizens to donate metals and conserve fibre, as a matter of significant patriotic importance. Resource conservation programmes established during the war were continued in some countries without an abundance of natural resources, such as Japan, after the war ended.

## Post-war Recycling

The next big investment in recycling occurred in the 1970s, due to rising energy costs. Recycling aluminium uses only five per cent of the energy required by virgin production; glass, paper and metals have less dramatic but very significant energy savings when recycled feedstock is used.

## Supply

For a recycling programme to work, having a large, stable supply of recyclable material is crucial. Three legislative options have been used to create such a supply: mandatory recycling collection, container deposit legislation, and refuse bans. Mandatory collection laws set recycling targets for cities to aim for, usually in the form that a certain percentage of a material must be diverted from the city's waste stream by a target date. The city is then responsible for working to meet this target.

Container deposit legislation involves offering a refund for the return of certain containers, typically glass, plastic, and metal. When a product in such a container is purchased, a small surcharge is added to the price. This surcharge can be reclaimed by the consumer if the container is returned to a collection point. These programmes have been very successful, often resulting in an 80 per cent recycling rate. Despite such good results, the shift in collection costs from local government to industry and consumers has created strong opposition to the creation of such programmes in some areas.

A third method of increase supply of recyclates is to ban the disposal of certain materials as waste, often including used oil, old batteries, tires and garden waste. One aim of this method is to create a viable economy for proper disposal of banned products. Care must be taken that enough of these recycling services exist, or such bans simply lead to increased illegal dumping.

Legislation has also been used to increase and maintain a demand for recycled materials. Four methods of such legislation exist: minimum recycled content mandates, utilization rates, procurement policies, recycled product labeling.

Both minimum recycled content mandates and utilization rates increase demand directly by forcing manufacturers to include recycling in their operations. Content mandates specify that a certain percentage of a new product must consist of recycled material. Utilization rates are a more flexible option: industries are permitted to meet the recycling targets at any point of their operation or even contract recycling out in exchange for tradeable credits. Opponents to both of these methods point to the large increase in reporting requirements they impose, and claim that they rob industry of necessary flexibility.

Governments have used their own purchasing power to increase recycling demand through what are called 'procurement policies'. These policies are either 'set-asides', which earmark a certain amount of spending solely towards recycled products, or 'price preference' programmes which provide a larger budget when recycled items are purchased. Additional regulations can target specific cases: in the United States, for example, the Environmental Protection Agency mandates the purchase of oil, paper, tires and building insulation from recycled or re-refined sources whenever possible.

The final government regulation towards increased demand is recycled product labeling. When producers are required to label their packaging with amount of recycled material in the product (including the packaging), consumers are better able to make educated choices. Consumers with sufficient buying power can then choose more environmentally conscious options, prompt producers to increase the amount of recycled material in their products, and indirectly increase demand. Standardized recycling labeling can also have a positive effect on supply of recyclates if the labeling includes information on how and where the product can be recycled.

A number of different systems have been implemented to collect recyclates from the general waste stream. These systems lie along the spectrum of trade-off between public convenience and government ease and expense. The three main categories of collection are 'drop-off centres', 'buy-back centres' and 'curbside collection'.

Drop-off centres require the waste producer to carry the recyclates to a central location, either an installed or mobile collection station or the reprocessing plant itself. They are the easiest type of collection to establish, but suffer from low and unpredictable throughput. Buy-back centres differ in that the cleaned recyclates are purchased, thus providing a clear incentive for use and creating a stable supply. The post-processed material can then be sold on, hopefully creating a profit. Unfortunately government subsidies are necessary to make buy-back centres a viable enterprise, as according to the United States Nation Solid Wastes Management Association it costs on average US$50 to process a ton of material, which can only be resold for US$30

Curbside collection encompasses many subtly different systems, which differ mostly on where in the process the recyclates are sorted and cleaned. The main categories are mixed waste collection, commingled recyclables and source separation. A waste collection vehicle generally picks up the waste.

A recycling truck collecting the contents of a recycling bin in Canberra, Australia.

At one end of the spectrum is mixed waste collection, in which all recyclates are collected mixed in with the rest of the waste, and the desired material is then sorted out and cleaned at a central sorting facility. This results in a large amount of recyclable waste, paper especially, being too soiled to reprocess, but has advantages as well: the city need not pay for a separate collection of recyclates and no public education is needed. Any changes to which materials are recyclable is easy to accommodate as all sorting happens in a central location.

In a Commingled or single-stream system, all recyclables for collection are mixed but kept separate from other waste. This greatly reduces the need for post-collection cleaning but does require public education on what materials are recyclable.

Source separation is the other extreme, where each material is cleaned and sorted prior to collection. This method requires the least post-collection sorting and produces the purest recyclates, but incurs additional operating costs for collection of each separate material. An extensive public education programme is also required, which must be successful if recyclate contamination is to be avoided.

Source separation used to be the preferred method due to the high sorting costs incurred by commingled collection. Advances in sorting technology , however, have lowered this overhead substantially—many areas which had developed source separation programmes have since switched to comingled collection.

Once commingled recyclates are collected and delivered to a central collection facility, the different types of materials must be sorted. This is done in a series of stages, many of which involve automated processes such that a truck-load of material can be fully sorted in less than an hour. Some plants can now sort the materials automatically, known as single-stream recycling. A 30 per cent increase in recycling rates has been seen in the areas where these plants exist.

Initially, the commingled recyclates are removed from the collection vehicle and placed on a conveyor belt spread out in a single layer. Large pieces of corrugated fiberboard and plastic bags are removed by hand at this stage, as they can cause later machinery to jam.

Next, automated machinery separates the recyclates by weight, splitting lighter paper and plastic from heavier glass and metal. Cardboard is removed from the mixed paper, and the most common types of plastic, PET (#1) and HDPE (#2), are collected. This separation is usually done by hand, but has become automated in some sorting centers: a spectroscopic scanner is used to differentiate between different types of paper and plastic based on the absorbed wavelengths, and subsequently divert each material into the proper collection channel.

Strong magnets are used to separate out ferrous metals, such as iron, steel, and tin-plated steel cans ("tin cans"). Non-ferrous metals are ejected by magnetic eddy currents in which a rotating magnetic field induces an electric current around the aluminium cans, which in turn creates a magnetic eddy current inside the cans. This magnetic eddy current is repulsed by a large magnetic field, and the cans are ejected from the rest of the recyclate stream.

Glass can be smashed up into little bits and put in a fire and melted to form new glass There is some debate over whether recycling is economically efficient. Municipalities often see fiscal benefits from implementing recycling programmes, largely due to the reduced landfill costs. A study conducted by the Technical University of Denmark found that in 83 per cent of cases, recycling is the most efficient method to dispose of household waste. However, a 2004 assessment by the Danish Environmental Assessment Institute concluded that incineration was the most effective method for disposing of drink containers, even aluminium ones.

Fiscal efficiency is separate from economic efficiency. Economic analysis of recycling includes what economists call externalities, which are unpriced costs and benefits that accrue to individuals outside of private transactions. Examples include: decreased air pollution and greenhouse gases from incineration, reduced hazardous waste leaching from landfills, reduced energy consumption, and reduced waste and resource consumption, which leads to a reduction in environmentally damaging mining and timber activity. About 4,000 minerals have been identified, of these around 100 can be called common, another several hundred are relatively common, and the rest are rare. Without more recycling, zinc could be used up by 2037, both indium and hafnium could run out by 2017, and terbium could be gone before 2012. At current rates, reserves of phosphorus will be depleted in the next 50 to 100 years. Without mechanisms such as taxes or subsidies to internalize externalities, businesses will ignore them despite the costs imposed on society. To make such non-fiscal benefits economically relevant, advocates have pushed for legislative action to increase the demand for recycled materials. The United States Environmental Protection Agency

(EPA) has concluded in favor of recycling, saying that recycling efforts reduced the country's carbon emissions by a net 49 million metric tonnes in 2005. In the United Kingdom, the Waste and Resources Action Programme stated that Great Britain's recycling efforts reduce $CO_2$ emissions by 10-15 million tonnes a year. Recycling is more efficient in densely populated areas, as there are economies of scale involved.

Certain requirements must be met for recycling to be economically feasible and environmentally effective. These include an adequate source of recyclates, a system to extract those recyclates from the waste stream, a nearby factory capable of reprocessing the recyclates, and a potential demand for the recycled products. These last two requirements are often overlooked—without both an industrial market for production using the collected materials and a consumer market for the manufactured goods, recycling is incomplete and in fact only 'collection'.

Many economists favour a moderate level of government intervention to provide recycling services. Economists of this mindset probably view product disposal as an externality of production and subsequently argue government is most capable of alleviating such a dilemma. However, those of the laissez faire approach to municipal recycling see product disposal as a service that consumers value. A free-market approach is more likely to suit the preferences of consumers since profit-seeking businesses have greater incentive to produce a quality product or service than does government. Moreover, economists most always advise against government intrusion in any market with little or no externalities'.

Certain countries trade in unprocessed recyclates. Some have complained that the ultimate fate of recyclates sold to another country is unknown and they may end up in landfills instead of reprocessed. According to one report, in America, 50-80 per cent of computers destined for recycling are actually not recycled. There are reports of illegal-waste imports to China being dismantled and recycled solely for monetary gain, without consideration for workers' health or environmental damage. Though the Chinese government has banned these practices, it has not been able to eradicate them. In 2008, the prices of recyclable waste plummeted before rebounding in 2009. Cardboard averaged about £53/tonne from 2004-2008, dropped to £19/tonne, and then went up to £59/tonne in May 2009. PET plastic averaged about £156/tonne, dropped to £75/tonne and then moved up to £195/tonne in May 2009. Certain regions have difficulty using or exporting as much of a material as they recycle. This problem is most prevalent with glass: both Britain and the U.S. import large quantities of wine bottled in green glass. Though much of this glass is sent to be recycled, outside the American Midwest there is not enough wine production to use all of the reprocessed material. The extra must be downcycled into building materials or re-inserted into the regular waste stream.

Similarly, the northwestern United States has difficulty finding markets for recycled newspaper, given the large number of pulp mills in the region as well as the proximity to Asian markets. In other areas of the U.S., however, demand for used newsprint has seen wide fluctuation.

In some U.S. states, a programme called RecycleBank pays people with coupons to recycle, receiving money from local municipalities for the reduction in landfill space which must be purchased. It uses a single stream process in which all material is automatically sorted.

Much of the difficulty inherent in recycling comes from the fact that most products are not designed with recycling in mind. The concept of sustainable design aims to solve this problem, and was laid out in the book "*Cradle to Cradle: Remaking the Way We Make Things*" by architect William McDonough and chemist Michael Braungart. They suggest that every product (and all packaging they require) should have a complete 'closed-loop' cycle mapped out for each component—a way in which every component will either return to the natural ecosystem through biodegradation or be recycled indefinitely.

As with environmental economics, care must be taken to ensure a complete view of the costs and benefits involved. For example, cardboard packaging for food products is more easily recycled than plastic, but is heavier to ship and may result in more waste from spoilage.

The following are criticisms of many popular points used for recycling.

### Saves Energy

There is controversy on just how much energy is saved through recycling. The Energy Information Administration (EIA) states on its website that "a paper mill uses 40 per cent less energy to make paper from recycled paper than it does to make paper from fresh lumber." Critics often argue that in the overall processes, it can take more energy to produce recycled products than it does to dispose of them in traditional landfill methods. This argument is followed from the curbside collection of recyclables, which critics note is often done by a second waste truck. Recycling proponents point out that a second timber or logging truck is eliminated when paper is collected for recycling.

It is difficult to determine the exact amount of energy consumed or produced in waste disposal processes. How much energy is used in recycling depends largely on the type of material being recycled and the process used to do so. Aluminium is generally agreed to use far less energy when recycled rather than being produced from scratch. The EPA states that "recycling aluminum cans, for example, saves 95 per cent of the energy required to make the same amount of aluminum from its virgin source, bauxite".

Economist Steven Landsburg has suggested that the sole benefit of reducing landfill space is trumped by the energy needed and resulting pollution from the recycling process. Others, however, have calculated through life cycle assessment that producing recycled paper uses less energy and water than harvesting, pulping, processing, and transporting virgin trees. When less recycled paper is used, additional energy is needed to create and maintain farmed forests until these forests are as self-sustainable as virgin forests.

Other studies have shown that recycling in itself is inefficient to perform the "decoupling" of economic development from the depletion of non-renewable raw materials that is necessary for sustainable development. When global consumption of a natural resource grows by more than one per cent per annum, its depletion is inevitable, and the best recycling can do is to delay it by a number of years. Nevertheless, if this decoupling can be achieved by other means, so that consumption of the resource is reduced below one per cent per annum, then recycling becomes indispensable — indeed recycling rates above 80 per cent are required for a significant slowdown of the resource depletion.

The amount of money actually saved through recycling depends on the efficiency of the recycling programme used to do it. The Institute for Local Self-reliance argues that the cost of recycling depends on various factors around a community that recycles, such as landfill fees and the amount of disposal that the community recycles. It states that communities start to save money when they treat recycling as a replacement for their traditional waste system rather than an add-on to it and by "redesigning their collection schedules and/or trucks."

In many cases, the cost of recyclable materials also exceeds the cost of raw materials. Virgin plastic resin costs 40 per cent less than recycled resin. Additionally, a United States Environmental Protection Agency (EPA) study that tracked the price of clear glass from July 15 to August 2, 1991, found that the average cost per ton ranged from $40 to $60, while a USGS report shows that the cost per ton of raw silica sand from years 1993 to 1997 fell between $17.33 and $18.10.

In a 1996 article for *The New York Times*, John Tierney argued that it costs more money to recycle the trash of New York City than it does to dispose of it in a landfill. Tierney argued that the recycling process employs people to do the additional waste disposal, sorting, inspecting, and many fees are often charged because the processing costs used to make the end product are often more than the profit from its sale. Tierney also referenced a study conducted by the Solid Waste Association of North America (SWANA) that found in the six communities involved in the study, "all but one of the curbside recycling programmes, and all the composting operations and waste-to-energy incinerators, increased the cost of waste disposal".

Tierney also points out that "the prices paid for scrap materials are a measure of their environmental value as recyclables. Scrap aluminum fetches a high price because recycling it consumes so much less energy than manufacturing new aluminum".

Critics often argue that while recycling may create jobs, they are often jobs with low wages and terrible working conditions. These jobs are sometimes considered to be make-work jobs that don't produce as much as the cost of wages to pay for those jobs. In areas without many environmental regulations and/or worker protections, jobs involved in recycling such as ship breaking can result in deplorable conditions for both workers and the surrounding communities.

Recycling proponents counter that the jobs involved in recovery of an equal amount of virgin material creates worse jobs. Timber harvesting and ore mining are more dangerous than paper recycling and metal recycling.

## Saves Trees

Economist Steven Landsburg, author of a paper entitled 'Why I Am Not an Environmentalist', has claimed that paper recycling actually reduces tree populations. He argues that because paper companies have incentives to replenish the forests they own, large demands for paper lead to large forests. Conversely, reduced demand for paper leads to fewer 'farmed' forests. Similar arguments were expressed in a 1995 article for The Free Market.

When foresting companies cut down trees, more are planted in their place. Most paper comes from pulp forests grown specifically for paper production. Many environmentalists point out, however, that 'farmed' forests are inferior to virgin forests in several ways. Farmed forests are not able to fix the soil as quickly as virgin forests, causing widespread soil erosion and often requiring large amounts of fertilizer to maintain while containing little tree and wild-life biodiversity compared to virgin forests. Also, the new trees planted are not as big as the trees that were cut down, and the argument that there will be 'more trees' is not compelling to forestry advocates when they are counting saplings.

The recycling of paper should not be confused with saving the tropical forest. Many people have the misconception that paper-making is what's causing deforestation of tropical rain forests but rarely any tropical wood is harvested for paper. Deforestation is mainly caused by population pressure such as demand of more land for agriculture or construction use. Therefore, the recycling paper, although reduces demand of trees, doesn't greatly benefit the tropical rain forests.

In some prosperous and many less prosperous countries in the world, the traditional job of recycling is performed by the entrepreneurial poor such as the karung guni, Zabaleen, the rag and bone man, waste picker, and junk man. With the creation of large recycling organizations that may be profitable, either by law or economies of scale, the poor are more likely to be driven out of the recycling and the remanufacturing market. To compensate for this loss of income to the poor, a society may need to create additional forms of societal programmes to help support the poor. Like the parable of the broken window, there is a net loss to the poor and possibly the whole of a society to make recycling artificially profitable through law. However, as seen in Brazil and Argentina, waste pickers/informal recyclers are able to work alongside governments, in (semi)funded cooperatives, allowing informal recycling to be legitimized as a paying government job.

Because the social support of a country is likely less than the loss of income to the poor doing recycling, there is a greater chance that the poor will come in conflict with the large recycling organizations. This means fewer people can decide if certain waste is more economically reusable in its current form rather than being reprocessed. Contrasted to the recycling poor, the efficiency of their recycling may actually be higher for some materials because individuals have greater control over what is considered 'waste'.

One labour-intensive underused waste is electronic and computer waste. Because this waste may still be functional and wanted mostly by the poor, the poor may sell or use it at a greater efficiency than large recyclers.

Many recycling advocates believe that this laissez-faire individual-based recycling does not cover all of society's recycling needs. Thus, it does not negate the need for an organized recycling programme. Local government often consider the activities of the recycling poor as contributing to property blight.

## Public Participation in Recycling Programmes

Many studies have addressed recycling behaviour and strategies to encourage community involvement in recycling programmes. It has been argued that recycling behaviour is not natural because it requires a focus and appreciation for long term planning, whereas humans have evolved to be sensitive to short term survival goals; and that to overcome this innate predisposition, the best solution would be to use social pressure to compel participation in recycling programmes. However, recent studies have concluded that social pressure is unviable in this context. One reason for this is that social pressure functions well in small group sizes of 50 to 150 indiviudals (common to nomadic hunter-gatherer peoples) but not in communities numbering in the millions, as we see today. Another reason is that individual recycling does not take place in the public view.

CHAPTER 12

# Air Pollution

Air pollution is the introduction of chemicals, particulate matter, or biological materials that cause harm or discomfort to humans or other living organisms, or damages the natural environment into the atmosphere.

The atmosphere is a complex dynamic natural gaseous system that is essential to support life on planet Earth. Stratospheric ozone depletion due to air pollution has long been recognized as a threat to human health as well as to the Earth's ecosystems.

Indoor air pollution and urban air quality are listed as two of the world's worst pollution problems in the 2008 Blacksmith Institute World's Worst Polluted Places report.

An air pollutant is known as a substance in the air that can cause harm to humans and the environment. Pollutants can be in the form of solid particles, liquid droplets, or gases. In addition, they may be natural or man-made.

Pollutants can be classified as either primary or secondary. Usually, primary pollutants are substances directly emitted from a process, such as ash from a volcanic eruption, the carbon monoxide gas from a motor vehicle exhaust or sulfur dioxide released from factories.

Secondary pollutants are not emitted directly. Rather, they form in the air when primary pollutants react or interact. An important example of a secondary pollutant is ground level ozone — one of the many secondary pollutants that make up photochemical smog.

Note that some pollutants may be both primary and secondary: that is, they are both emitted directly and formed from other primary pollutants.

About four per cent of deaths in the United States can be attributed to air pollution, according to the Environmental Science Engineering Programme at the Harvard School of Public Health.

Major primary pollutants produced by human activity include:

- *Sulfur Oxides ($SO_x$)* — especially sulfur dioxide, a chemical compound with the formula $SO_2$. $SO_2$ is produced by volcanoes and in various industrial processes. Since coal and petroleum often contain sulfur compounds, their combustion generates sulfur dioxide. Further oxidation of $SO_2$, usually in the presence of a catalyst such as $NO_2$, forms $H_2SO_4$, and thus acid rain. This is one of the causes for concern over the environmental impact of the use of these fuels as power sources.
- *Nitrogen Oxides ($NO_x$)* — especially nitrogen dioxide are emitted from high temperature combustion. Can be seen as the brown haze dome above or plume downwind of cities. Nitrogen dioxide is the chemical compound with the formula $NO_2$. It is one of the several nitrogen oxides. This reddish-brown toxic gas has a characteristic sharp, biting odor. $NO_2$ is one of the most prominent air pollutants.
- *Carbon Monoxide* — is a colourless, odourless, non-irritating but very poisonous gas. It is a product by incomplete combustion of fuel such as natural gas, coal or wood. Vehicular exhaust is a major source of carbon monoxide.
- *Carbon Dioxide ($CO_2$)* — a greenhouse gas emitted from combustion but is also a gas vital to living organisms. It is a natural gas in the atmosphere.
- *Volatile Organic Compounds* — VOCs are an important outdoor air pollutant. In this field they are often divided into the separate categories of methane ($CH_4$) and non-methane (NMVOCs). Methane is an extremely efficient greenhouse gas which contributes to enhanced global warming. Other hydrocarbon VOCs are also significant greenhouse gases via their role in creating ozone and in prolonging the life of methane in the atmosphere, although the effect varies depending on local air quality. Within the NMVOCs, the aromatic compounds benzene, toluene and xylene are suspected carcinogens and may lead to leukemia through prolonged exposure. 1,3-butadiene is another dangerous compound which is often associated with industrial uses.
- *Particulate Matter* — Particulates, alternatively referred to as particulate matter (PM) or fine particles, are tiny particles of solid or

liquid suspended in a gas. In contrast, aerosol refers to particles and the gas together. Sources of particulate matter can be man made or natural. Some particulates occur naturally, originating from volcanoes, dust storms, forest and grassland fires, living vegetation, and sea spray. Human activities, such as the burning of fossil fuels in vehicles, power plants and various industrial processes also generate significant amounts of aerosols. Averaged over the globe, anthropogenic aerosols–those made by human activities—currently account for about 10 per cent of the total amount of aerosols in our atmosphere. Increased levels of fine particles in the air are linked to health hazards such as heart disease, altered lung function and lung cancer.

- *Persistent free radicals* connected to airborne fine particles could cause cardiopulmonary disease.
- *Toxic metals,* such as lead, cadmium and copper.
- *Chlorofluorocarbons (CFCs)* — harmful to the ozone layer emitted from products currently banned from use.
- *Ammonia* ($NH_3$) — emitted from agricultural processes. Ammonia is a compound with the formula $NH_3$. It is normally encountered as a gas with a characteristic pungent odor. Ammonia contributes significantly to the nutritional needs of terrestrial organisms by serving as a precursor to foodstuffs and fertilizers. Ammonia, either directly or indirectly, is also a building block for the synthesis of many pharmaceuticals. Although in wide use, ammonia is both caustic and hazardous.
- *Odours* — such as from garbage, sewage, and industrial processes.
- *Radioactive Pollutants* — produced by nuclear explosions, war explosives, and natural processes such as the radioactive decay of radon.

Secondary pollutants include:

- Particulate matter formed from gaseous primary pollutants and compounds in photochemical smog. Smog is a kind of air pollution; the word 'smog' is a portmanteau of smoke and fog. Classic smog results from large amounts of coal burning in an area caused by a mixture of smoke and sulfur dioxide. Modern smog does not usually come from coal but from vehicular and industrial emissions that are acted on in the atmosphere by sunlight to form secondary pollutants that also combine with the primary emissions to form photochemical smog.
- *Ground level ozone* ($O_3$) formed from $NO_x$ and VOCs. Ozone ($O_3$) is a key constituent of the troposphere (it is also an important constituent of certain regions of the stratosphere commonly known as the Ozone

layer). Photochemical and chemical reactions involving it drive many of the chemical processes that occur in the atmosphere by day and by night. At abnormally high concentrations brought about by human activities (largely the combustion of fossil fuel), it is a pollutant, and a constituent of smog.

- *Peroxyacetyl Nitrate* (PAN) — similarly formed from $NO_x$ and VOCs.

Minor air pollutants include:

- A large number of minor hazardous air pollutants. Some of these are regulated in USA under the Clean Air Act and in Europe under the Air Framework Directive.
- A variety of persistent organic pollutants, which can attach to particulate matter.

Persistent organic pollutants (POPs) are organic compounds that are resistant to environmental degradation through chemical, biological, and photolytic processes. Because of this, they have been observed to persist in the environment, to be capable of long-range transport, bioaccumulate in human and animal tissue, biomagnify in food chains, and to have potential significant impacts on human health and the environment.

Sources of air pollution refer to the various locations, activities or factors which are responsible for the releasing of pollutants in the atmosphere. These sources can be classified into two major categories which are:

- *Anthropogenic sources* (human activity) mostly related to burning different kinds of fuel;
- *'Stationary Sources'* include smoke stacks of power plants, manufacturing facilities (factories) and waste incinerators, as well as furnaces and other types of fuel-burning heating devices.

'Mobile Sources' include motor vehicles, marine vessels, aircraft and the effect of sound etc.

Chemicals, dust and controlled burn practices in agriculture and forestry management. Controlled or prescribed burning is a technique sometimes used in forest management, farming, prairie restoration or greenhouse gas abatement. Fire is a natural part of both forest and grassland ecology and controlled fire can be a tool for foresters. Controlled burning stimulates the germination of some desirable forest trees, thus renewing the forest.

Fumes from paint, hair spray, varnish, aerosol sprays and other solvents.

Waste deposition in landfills, which generate methane.Methane is not toxic; however, it is highly flammable and may form explosive mixtures with air. Methane is also an asphyxiant and may displace oxygen in an enclosed space. Asphyxia or suffocation may result if the oxygen concentration is reduced to below 19.5 per cent by displacement.

Military, such as nuclear weapons, toxic gases, germ warfare and rocketry.

**Natural Sources**

- Dust from natural sources, usually large areas of land with little or no vegetation.
- Methane, emitted by the digestion of food by animals, for example cattle.
- Radon gas from radioactive decay within the Earth's crust. Radon is a colourless, odorless, naturally occurring, radioactive noble gas that is formed from the decay of radium. It is considered to be a health hazard. Radon gas from natural sources can accumulate in buildings, especially in confined areas such as the basement and it is the second most frequent cause of lung cancer, after cigarette smoking.
- Smoke and carbon monoxide from wildfires.
- Volcanic activity, which produce sulfur, chlorine, and ash particulates.

**Emission Factors**

Air pollutant emission factors are representative values that people attempt to relate the quantity of a pollutant released to the ambient air with an activity associated with the release of that pollutant. These factors are usually expressed as the weight of pollutant divided by a unit weight, volume, distance, or duration of the activity emitting the pollutant (e.g., kilograms of particulate emitted per megagram of coal burned). Such factors facilitate estimation of emissions from various sources of air pollution. In most cases, these factors are simply averages of all available data of acceptable quality, and are generally assumed to be representative of long-term averages.

The United States Environmental Protection Agency has published a compilation of air pollutant emission factors for a multitude of industrial sources. The United Kingdom, Australia, Canada and many other countries have published similar compilations, as well as the European Environment Agency.

**Indoor Air Quality (IAQ)**

A lack of ventilation indoors concentrates air pollution where people often spend the majority of their time. Radon (Rn) gas, a carcinogen, is exuded from the Earth in certain locations and trapped inside houses. Building materials including carpeting and plywood emit formaldehyde ($H_2CO$) gas. Paint and solvents give off volatile organic compounds (VOCs) as they dry. Lead paint can degenerate into dust and be inhaled. Intentional air pollution is introduced with the use of air fresheners, incense, and other scented items. Controlled wood fires in stoves and fireplaces can add significant amounts of smoke particulates into the air, inside and out. Indoor pollution fatalities may be caused by using pesticides and other chemical sprays indoors without proper ventilation.

Carbon monoxide (CO) poisoning and fatalities are often caused by faulty vents and chimneys, or by the burning of charcoal indoors. Chronic carbon monoxide poisoning can result even from poorly adjusted pilot lights. Traps are built into all domestic plumbing to keep sewer gas, hydrogen sulfide, out of interiors. Clothing emits tetrachloroethylene, or other dry cleaning fluids, for days after dry cleaning.

Though its use has now been banned in many countries, the extensive use of asbestos in industrial and domestic environments in the past has left a potentially very dangerous material in many localities. Asbestosis is a chronic inflammatory medical condition affecting the tissue of the lungs. It occurs after long-term, heavy exposure to asbestos from asbestos-containing materials in structures. Sufferers have severe dyspnea (shortness of breath) and are at an increased risk regarding several different types of lung cancer. As clear explanations are not always stressed in non-technical literature, care should be taken to distinguish between several forms of relevant diseases. According to the World Health Organisation (WHO) , these may defined as; asbestosis, *lung cancer*, and *mesothelioma* (generally a very rare form of cancer, when more widespread it is almost always associated with prolonged exposure to asbestos).

Biological sources of air pollution are also found indoors, as gases and airborne particulates. Pets produce dander, people produce dust from minute skin flakes and decomposed hair, dust mites in bedding, carpeting and furniture produce enzymes and micrometre-sized fecal droppings, inhabitants emit methane, mold forms in walls and generates mycotoxins and spores, air conditioning systems can incubate Legionnaires' disease and mold, and houseplants, soil and surrounding gardens can produce pollen, dust, and mold. Indoors, the lack of air circulation allows these airborne pollutants to accumulate more than they would otherwise occur in nature.

The World Health Organization states that 2.4 million people die each year from chronic obstructive pulmonary disease (COPD), with 700 000 of these deaths attributable to indoor air pollution. "Epidemiological studies suggest that more than 500,000 Americans die each year from cardiopulmonary disease linked to breathing fine particle air pollution. . ." A study by the University of Birmingham has shown a strong correlation between pneumonia related deaths and air pollution from motor vehicles. Worldwide more deaths per year are linked to air pollution than to automobile accidents. Published in 2005 suggests that 310,000 Europeans die from air pollution annually. Causes of deaths include aggravated asthma, emphysema, lung and heart diseases, and respiratory allergies. The US EPA estimates that a proposed set of changes in diesel engine technology (*Tier 2*) could result in 12,000 fewer *premature mortalities*, 15,000 fewer heart attacks, 6,000 fewer emergency room visits by children with asthma, and 8,900 fewer respiratory-related hospital admissions each year in the United States.

The worst short term civilian pollution crisis in India was the 1984 Bhopal Disaster. Leaked industrial vapors from the Union Carbide factory, belonging to Union Carbide, Inc., U.S.A., killed more than 25,000 people outright and injured anywhere from 150,000 to 600,000. The United Kingdom suffered its worst air pollution event when the December 4 Great Smog of 1952 formed over London. In six days more than 4,000 died, and 8,000 more died within the following months. An accidental leak of anthrax spores from a biological warfare laboratory in the former USSR in 1979 near Sverdlovsk is believed to have been the cause of hundreds of civilian deaths. The worst single incident of air pollution to occur in the United States of America occurred in Donora, Pennsylvania in late October, 1948, when 20 people died and over 7,000 were injured.

The health effects caused by air pollutants may include difficulty in breathing, wheezing, coughing and aggravation of existing respiratory and cardiac conditions. These effects can result in increased medication use, increased doctor or emergency room visits, more hospital admissions and premature death. The human health effects of poor air quality are far reaching, but principally affect the body's respiratory system and the cardiovascular system. Individual reactions to air pollutants depend on the type of pollutant a person is exposed to, the degree of exposure, the individual's health status and genetics.

A new economic study of the health impacts and associated costs of air pollution in the Los Angeles Basin and San Joaquin Valley of Southern California shows that more than 3800 people die prematurely (approximately 14 years earlier than normal) each year because air pollution levels violate federal standards. The number of annual premature deaths is considerably higher than the fatalities related to auto collisions in the same area, which average fewer than 2,000 per year.

Diesel exhaust (DE) is a major contributor to combustion derived particulate matter air pollution. In several human experimental studies, using a well validated exposure chamber setup, DE has been linked to acute vascular dysfunction and increased thrombus formation. This serves as a plausible mechanistic link between the previously described association between particulate matter air pollution and increased cardiovascular morbidity and mortality.

A study from around the years of 1999 to 2000, by the University of Washington, showed that patients near and around particulate matter air pollution had an increased risk of pulmonary exacerbations and decrease in lung function. Patients were examined before the study for amounts of specific pollutants like *Pseudomonas aeruginosa* or *Burkholderia cenocepacia* as well as their socioeconomic standing. Participants involved in the study were located in the United States in close proximity to an Environmental Protection Agency. During the time of the study 117 deaths were associated with air pollution. Many patients in the study lived in or near large metropolitan areas in order to be close to medical help. These same patients had higher level of pollutants found in their system because of more emissions in larger cities. As cystic fibrosis patients already suffer from decreased lung function, everyday pollutants such as smoke, emissions from automobiles, tobacco smoke and improper use of indoor heating devices could further compromise lung function.

## Effects on COPD

Chronic obstructive pulmonary disease (COPD) include diseases such as chronic bronchitis, emphysema, and some forms of asthma.

A study conducted in 1960-1961 in the wake of the Great Smog of 1952 compared 293 London residents with 477 residents of Gloucester, Peterborough, and Norwich, three towns with low reported death rates from chronic bronchitis. All subjects were male postal truck drivers aged 40 to 59. Compared to the subjects from the outlying towns, the London subjects exhibited more severe respiratory symptoms (including cough, phlegm, and dyspnea), reduced lung function ($FEV_1$ and peak flow rate), and increased sputum production and purulence. The differences were more pronounced for subjects aged 50 to 59. The study controlled for age and smoking habits, so concluded that air pollution was the most likely cause of the observed differences.

It is believed that much like cystic fibrosis, by living in a more urban environment serious health hazards become more apparent. Studies have shown that in urban areas patients suffer mucus hypersecretion, lower levels of lung function, and more self diagnosis of chronic bronchitis and emphysema.

Cities around the world with high exposure to air pollutants have the possibility of children living within them to develop asthma, pneumonia and other lower respiratory infections as well as a low initial birth rate. Protective measures to ensure the youths' health are being taken in cities such as New Delhi, India where buses now use compressed natural gas to help eliminate the 'pea-soup' smog. Research by the World Health Organization shows there is the greatest concentration of particulate matter particles in countries with low economic world power and high poverty and population rates. Examples of these countries include Egypt, Sudan, Mongolia, and Indonesia. The Clean Air Act was passed in 1970, however in 2002 at least 146 million Americans were living in areas that did not meet at least one of the 'criteria pollutants' laid out in the 1997 National Ambient Air Quality Standards. Those pollutants included: ozone, particulate matter, sulfur dioxide, nitrogen dioxide, carbon monoxide, and lead. Because children are outdoors more and have higher minute ventilation they are more susceptible to the dangers of air pollution.

## Health Effects in Relatively 'Clean' Areas

Even in areas with relatively low levels of air pollution, public health effects can be substantial and costly. This is because effects can occur at very low levels and a large number of people can potentially breathe in such pollutants. A 2005 scientific study for the British Columbia Lung Association showed that a one per cent improvement in ambient PM2.5 and ozone concentrations will produce a $29 million in annual savings in the region in 2010. This finding is based on health valuation of lethal (mortality) and sub-lethal (morbidity) effects.

## Reduction Efforts

There are various air pollution control technologies and land use planning strategies available to reduce air pollution. At its most basic level land use planning is likely to involve zoning and transport infrastructure planning. In most developed countries, land use planning is an important part of social policy, ensuring that land is used efficiently for the benefit of the wider economy and population as well as to protect the environment.

Efforts to reduce pollution from mobile sources includes primary regulation (many developing countries have permissive regulations), expanding regulation to new sources (such as cruise and transport ships, farm equipment, and small gas-powered equipment such as lawn trimmers, chainsaws, and snowmobiles), increased fuel efficiency (such as through the use of hybrid vehicles), conversion to cleaner fuels (such as bioethanol, biodiesel, or conversion to electric vehicles).

## Control Devices

The following items are commonly used as pollution control devices by industry or transportation devices. They can either destroy contaminants or remove them from an exhaust stream before it is emitted into the atmosphere.

## Particulate Control

- *Mechanical collectors* ( dust cyclones, multicyclones)
- *Electrostatic precipitators* — An electrostatic precipitator (ESP), or electrostatic air cleaner is a particulate collection device that removes particles from a flowing gas (such as air) using the force of an induced electrostatic charge. Electrostatic precipitators are highly efficient filtration devices that minimally impede the flow of gases through the device, and can easily remove fine particulate matter such as dust and smoke from the air stream.
- *Baghouses* — Designed to handle heavy dust loads, a dust collector consists of a blower, dust filter, a filter-cleaning system, and a dust receptacle or dust removal system (distinguished from air cleaners which utilize disposable filters to remove the dust).
- *Particulate scrubbers* — Wet scrubber is a form of pollution control technology. The term describes a variety of devices that use pollutants from a furnace flue gas or from other gas streams. In a wet scrubber, the polluted gas stream is brought into contact with the scrubbing liquid, by spraying it with the liquid, by forcing it through a pool of liquid, or by some other contact method, so as to remove the pollutants.

In general, there are two types of air quality standards. *The first class of standards* (such as the U.S. National Ambient Air Quality Standards) set maximum atmospheric concentrations for specific pollutants. Environmental agencies enact regulations which are intended to result in attainment of these target levels. *The second class* (such as the North American Air Quality Index) take the form of a scale with various thresholds, which is used to communicate to the public the relative risk of outdoor activity. The scale may or may not distinguish between different pollutants.

## Cities

Air pollution is usually concentrated in densely populated metropolitan areas, especially in developing countries where environmental regulations are relatively lax or nonexistent. However, even populated areas in developed countries attain unhealthy levels of pollution.

The basic technology for analyzing air pollution is through the use of a variety of mathematical models for predicting the transport of air pollutants in the lower atmosphere. The principal methodologies are:

- *Point source dispersion*, used for industrial sources.
- *Line source dispersion*, used for airport and roadway air dispersion modelling.
- *Area source dispersion*, used for forest fires or duststorms.
- *Photochemical models*, used to analyze reactive pollutants that form smog.

The point source problem is the best understood, since it involves simpler mathematics and has been studied for a long period of time, dating back to about the year 1900. It uses a Gaussian dispersion model for buoyant pollution plumes to forecast the air pollution isopleths, with consideration given to wind velocity, stack height, emission rate and stability class (a measure of atmospheric turbulence). This model has been extensively validated and calibrated with experimental data for all sorts of atmospheric conditions.

The roadway air dispersion model was developed starting in the late 1950s and early 1960s in response to requirements of the National Environmental Policy Act and the U.S. Department of Transportation (then known as the Federal Highway Administration) to understand impacts of proposed new highways upon air quality, especially in urban areas. Several research groups were active in this model development, among which were: the Environmental Research and Technology (ERT) group in Lexington, Massachusetts, the ESL Inc. group in Sunnyvale, California and the California Air Resources Board group in Sacramento, California. The research of the ESL group received a boost with a contract award from the United States Environmental Protection Agency to validate a line source model using sulfur hexafluoride as a tracer gas. This programme was successful in validating the line source model developed by ESL inc. Some of the earliest uses of the model were in court cases involving highway air pollution, the Arlington, Virginia portion of Interstate 66 and the New Jersey Turnpike widening project through East Brunswick, New Jersey.

Area source models were developed in 1971 through 1974 by the ERT and ESL groups, but addressed a smaller fraction of total air pollution emissions, so that their use and need was not as widespread as the line source model, which enjoyed hundreds of different applications as early as the 1970s. Similarly photochemical models were developed primarily in the 1960s and 1970s, but their use was more specialized and for regional needs, such as understanding smog formation in Los Angeles, California.

## Environmental Impacts of *Greenhouse Gas* Pollutants

The greenhouse effect is a phenomenon whereby greenhouse gases create a condition in the upper atmosphere causing a trapping of heat and leading to increased surface and lower tropospheric temperatures. Carbon dioxide from combustion of fossil fuels is the major problem. Other greenhouse gases include methane, hydrofluorocarbons, perfluorocarbons, chlorofluorocarbons, nitrogen oxides, and ozone.

This effect has been understood by scientists for about a century, and technological advancements during this period have helped increase the breadth and depth of data relating to the phenomenon. Currently, scientists are studying the role of changes in composition of greenhouse gases from natural and anthropogenic sources for the effect on climate change.

A number of studies have also investigated the potential for long-term rising levels of atmospheric carbon dioxide to cause increases in the acidity of ocean waters and the possible effects of this on marine ecosystems.

### Acid Rain

Acid rain is a rain or any other form of precipitation that is unusually acidic, i.e. elevated levels of hydrogen ions (low pH). It can have harmful effects on plants, aquatic animals, and infrastructure through the process of wet deposition. Acid rain is caused by emissions of sulphur dioxide and nitrogen oxides which react with the water molecules in the atmosphere to produce acids. Governments have made efforts since the 1970s to reduce the release of sulfur dioxide into the atmosphere with positive results. However, it can also be caused naturally by the splitting of nitrogen compounds by the energy produced by lightning strikes, or the release of sulfur dioxide into the atmosphere by volcano eruptions.

'Acid rain' is a popular term referring to the deposition of wet (rain, snow, sleet, fog, cloudwater, and dew) and dry (acidifying particles and gases) acidic components. A more accurate term is 'acid deposition'. Distilled water, once carbon dioxide is removed, has a neutral pH of 7. Liquids with a pH less than 7 are acidic, and those with a pH greater than 7 are Alkaline. 'Clean' or unpolluted rain has a slightly acidic pH of over 5.7, because carbon dioxide and water in the air react together to form carbonic acid, but unpolluted rain also contains other chemicals.

$$H_2O\ (l) + CO_2\ (g) = H_2CO_3\ (aq)$$

Carbonic acid then can ionize in water forming low concentrations of hydronium and carbonate ions:

$$H_2O\ (l) + H_2CO_3\ (aq) = HCO_3^-\ (aq) + H_3O^+\ (aq)$$

Acid deposition as an environmental issue would include additional acids to $H_2CO_3$.

The corrosive effect of polluted, acidic city air on limestone and marble was noted in the 17th century by John Evelyn, who remarked upon the Arundel marbles "miserably neglected, & scattredup & downe about the Gardens & other places of Arundell-house, & how exceedingly the corrosive aire of *London* impaired them". Since the Industrial Revolution, emissions of sulfur dioxide and nitrogen oxides to the atmosphere have increased. In 1852, Robert Angus Smith was the first to show the relationship between acid rain and atmospheric pollution in Manchester, England. Though acidic rain was discovered in 1852, it was not until the late 1960s that scientists began widely observing and studying the phenomenon. The term 'acid rain' was generated in 1972. Canadian Harold Harvey was among the first to research a 'dead' lake. Public awareness of acid rain in the U.S increased in the 1970s after The New York Times promulgated reports from the Hubbard Brook Experimental Forest in New Hampshire of the myriad deleterious environmental effects demonstrated to result from it.

Occasional pH readings in rain and fog water of well below 2.4 have been reported in industrialized areas. Industrial acid rain is a substantial problem in China and Russia and areas down-wind from them. These areas all burn sulfur-containing coal to generate heat and electricity. The problem of acid rain not only has increased with population and industrial growth, but has become more widespread. The use of tall smokestacks to reduce local pollution has contributed to the spread of acid rain by releasing gases into regional atmospheric circulation. Often deposition occurs a considerable distance downwind of the emissions, with mountainous regions tending to receive the greatest deposition (simply because of their higher rainfall). An example of this effect is the low pH of rain (compared to the local emissions) which falls in Scandinavia.

In 1980, the U.S. Congress passed an Acid Deposition Act. This Act established a 10-year research programme under the direction of the National Acidic Precipitation Assessment Programme (NAPAP). NAPAP looked at the entire problem. It enlarged a network of monitoring sites to determine how acidic the precipitation actually was, and to determine long term trends, and established a network for dry deposition. It looked at the effects of acid rain and funded research on the effects of acid precipitation on freshwater and terrestrial ecosystems, historical buildings, monuments, and building materials. It also funded extensive studies on atmospheric processes and potential control programmes.

In 1991, NAPAP provided its first assessment of acid rain in the United States. It reported that five per cent of New England Lakes were acidic, with sulfates being the most common problem. They noted that two per cent of the lakes could no longer support Brook Trout, and six per cent of the lakes were unsuitable for the survival of many species of minnow. Subsequent Reports to Congress have documented chemical changes in soil and freshwater ecosystems, nitrogen saturation, decreases in amounts of nutrients in soil, episodic acidification, regional haze, and damage to historical monuments.

Meanwhile, in 1990, the US Congress passed a series of amendments to the Clean Air Act. Title IV of these amendments established the Acid Rain Programme, a cap and trade system designed to control emissions of sulfur dioxide and nitrogen oxides. Title IV called for a total reduction of about 10 million tons of $SO_2$ emissions from power plants. It was implemented in two phases. Phase I began in 1995, and limited sulfur dioxide emissions from 110 of the largest power plants to a combined total of 8.7 million tons of sulfur dioxide. One power plant in New England (Merrimack) was in Phase I. Four other plants (Newington, Mount Tom, Brayton Point, and Salem Harbor) were added under other provisions of the programme. Phase II began in 2000, and affects most of the power plants in the country.

During the 1990s, research has continued. On March 10, 2005, EPA issued the Clean Air Interstate Rule (CAIR). This rule provides states with a solution to the problem of power plant pollution that drifts from one state to another. CAIR will permanently cap emissions of $SO_2$ and NOx in the eastern United States. When fully implemented, CAIR will reduce $SO_2$ emissions in 28 eastern states and the District of Columbia by over 70 per cent and $NO_x$ emissions by over 60 per cent from 2003 levels.

Overall, the Programme's cap and trade programme has been successful in achieving its goals. Since the 1990s, $SO_2$ emissions have dropped 40 per cent, and according to the Pacific Research Institute, acid rain levels have dropped 65 per cent since 1976. However, this was significantly less successful than conventional regulation in the European Union, which saw a decrease of over 70 per cent in $SO_2$ emissions during the same time period.

In 2007, total $SO_2$ emissions were 8.9 million tons, achieving the programme's long term goal ahead of the 2010 statutory deadline.

The EPA estimates that by 2010, the overall costs of complying with the programme for businesses and consumers will be $1 billion to $2 billion a year, only one fourth of what was originally predicted.

## Emissions of Chemicals Leading to Acidification

The most important gas which leads to acidification is sulfur dioxide. Emissions of nitrogen oxides which are oxidized to form nitric acid are of increasing importance due to stricter controls on emissions of sulfur containing compounds. 70 Tg(S) per year in the form of $SO_2$ comes from fossil fuel combustion and industry, 2.8 Tg(S) from wildfires and 7-8 Tg(S) per year from volcanoes.

## Natural Phenomena

The principal natural phenomena that contribute acid-producing gases to the atmosphere are emissions from volcanoes. Thus, for example, fumaroles from Laguna Caliente crater of Poás Volcano create extremely high amounts of acid rain and fog with acidity 2 of pH, clearing an area of any vegetation and frequently causing irritation to the eyes and lungs of inhabitants in nearby settlements. Acid-producing gasses are created also by biological processes that occur on the land, in wetlands, and in the oceans. The major biological source of sulfur containing compounds is dimethyl sulfide.

Nitric acid in rainwater is an important source of fixed nitrogen for plant life, and is also produced by electrical activity in the atmosphere such as lightning.

Acidic deposits have been detected in glacial ice thousands of years old in remote parts of the globe.

Soils of Coniferous forests are naturally very acidic due to the shedding of needles and this phenomenon should not be confused with acid rain.

The principal cause of acid rain is sulfur and nitrogen compounds from human sources, such as electricity generation, factories, and motor vehicles. Coal power plants are one of the most polluting. The gases can be carried hundreds of kilometers in the atmosphere before they are converted to acids and deposited. In the past, factories had short funnels to let out smoke but this caused many problems locally; thus, factories now have taller smoke funnels. However, dispersal from these taller stacks causes pollutants to be carried farther, causing widespread ecological damage. Also, livestock production plays a major role. It is responsible for almost two-thirds of all anthropogenic sources of ammonia produced through human activities, which contributes significantly to acid rain.

## Oxidation

There are a large number of aqueous reactions that oxidize sulfur from S(IV) to S(VI), leading to the formation of sulfuric acid. The most important oxidation reactions are with ozone, hydrogen peroxide and oxygen (reactions with oxygen are catalyzed by iron and manganese in the cloud droplets).

## ACID DEPOSITION

### Wet Deposition

Wet deposition of acids occurs when any form of precipitation (rain, snow, etc.) removes acids from the atmosphere and delivers it to the Earth's surface. This can result from the deposition of acids produced in the raindrops (see aqueous phase chemistry above) or by the precipitation removing the acids either in clouds or below clouds. Wet removal of both gases and aerosols are both of importance for wet deposition.

### Dry Deposition

Acid deposition also occurs via dry deposition in the absence of precipitation. This can be responsible for as much as 20 to 60 per cent of total acid deposition. This occurs when particles and gases stick to the ground, plants or other surfaces.

### Adverse Effects

Acid rain has been shown to have adverse impacts on forests, freshwaters and soils, killing insect and aquatic life-forms as well as causing damage to buildings and having impacts on human health.

### Surface Waters and Aquatic Animals

Both the lower pH and higher aluminium concentrations in surface water that occur as a result of acid rain can cause damage to fish and other aquatic animals. At pHs lower than 5 most fish eggs will not hatch and lower pHs can kill adult fish. As lakes and rivers become more acidic biodiversity is reduced. Acid rain has eliminated insect life and some fish species, including the brook trout in some lakes, streams, and creeks in geographically sensitive areas, such as the Adirondack Mountains of the United States. However, the extent to which acid rain contributes directly or indirectly via runoff from the catchment to lake and river acidity (i.e., depending on characteristics of the surrounding watershed) is variable. The United States Environmental Protection Agency's (EPA) website states: "Of the lakes and streams surveyed, acid rain caused acidity in 75 per cent of the acidic lakes and about 50 per cent of the acidic streams".

Soil biology and chemistry can be seriously damaged by acid rain. Some microbes are unable to tolerate changes to low pHs and are killed. The enzymes of these microbes are denatured (changed in shape so they no longer function) by the acid. The hydronium ions of acid rain also mobilize toxins such as aluminium, and leach away essential nutrients and minerals such as magnesium.

$$2\,H^+\,(aq) + Mg^{2+}\,(clay) = 2\,H^+\,(clay) + Mg^{2+}\,(aq)$$

Soil chemistry can be dramatically changed when base cations, such as calcium and magnesium, are leached by acid rain thereby affecting sensitive species, such as sugar maple (Acer saccharum).

### Forests and Other Vegetation

Adverse effects may be indirectly related to acid rain, like the acid's effects on soil (see above) or high concentration of gaseous precursors to acid rain. High altitude forests are especially vulnerable as they are often surrounded by clouds and fog which are more acidic than rain.

Other plants can also be damaged by acid rain, but the effect on food crops is minimized by the application of lime and fertilizers to replace lost nutrients. In cultivated areas, limestone may also be added to increase the ability of the soil to keep the pH stable, but this tactic is largely unusable in the case of wilderness lands. When calcium is leached from the needles of red spruce, these trees become less cold tolerant and exhibit winter injury and even death.

### Human Health

Scientists have suggested direct links to human health. Fine particles, a large fraction of which are formed from the same gases as acid rain (sulfur dioxide and nitrogen dioxide), have been shown to cause premature deaths and illnesses such as cancer and other diseases. For more information on the health effects of aerosols see particulate health effects.

### Other Adverse Effects

Acid rain can also damage buildings and historic monuments, especially those made of rocks such as limestone and marble containing large amounts of calcium carbonate. Acids in the rain react with the calcium compounds in the stones to create gypsum, which then flakes off.

### Affected Areas

Places with significant impact by acid rain around the globe include most of eastern Europe from Poland northward into Scandinavia, the eastern third of the United States, and South Eastern Canada. Other affected areas include the South Eastern coast of China and Taiwan.

### Potential Problem Areas in the Future

Places like much of Southeast Asia (Indonesia, Malaysia and Thailand), Western South Africa (the country), Southern India and Sri Lanka and even West Africa (countries like Ghana, Togo and Nigeria) could all be prone to acidic rainfall in the future. As well as the many acids which are present in the atmosphere there are many in the soil.

## PREVENTION METHODS

### Technical Solutions

In the United States, many coal-burning power plants use Flue gas desulfurization (FGD) to remove sulfur-containing gases from their stack gases. An example of FGD is the wet scrubber which is commonly used in the U.S. and many other countries. A wet scrubber is basically a reaction tower equipped with a fan that extracts hot smoke stack gases from a power plant into the tower. Lime or limestone in slurry form is also injected into the tower to mix with the stack gases and combine with the sulfur dioxide present. The calcium carbonate of the limestone produces pH-neutral calcium sulfate that is physically removed from the scrubber. That is, the scrubber turns sulfur pollution into industrial sulfates.

In some areas the sulfates are sold to chemical companies as gypsum when the purity of calcium sulfate is high. In others, they are placed in landfill. However, the effects of acid rain can last for generations, as the effects of pH level change can stimulate the continued leaching of undesirable chemicals into otherwise pristine water sources, killing off vulnerable insect and fish species and blocking efforts to restore native life.

Vehicle emissions control reduces emissions of nitrogen oxides from motor vehicles.

### International Treaties

A number of international treaties on the long range transport of atmospheric pollutants have been agreed e.g. Sulphur Emissions Reduction Protocol under the Convention on Long-Range Transboundary Air Pollution.

### Emissions Trading

In this regulatory scheme, every current polluting facility is given or may purchase on an open market an emissions allowance for each unit of a designated pollutant it emits. Operators can then install pollution control equipment, and sell portions of their emissions allowances they no longer need for their own operations, thereby recovering some of the capital cost of their investment in such equipment. The intention is to give operators economic incentives to install pollution controls.

The first emissions trading market was established in the United States by enactment of the Clean Air Act Amendments of 1990. The overall goal of the Acid Rain Programme established by the Act is to achieve significant environmental and public health benefits through reductions in emissions of sulfur dioxide ($SO_2$) and nitrogen oxides ($NO_x$), the primary causes of acid rain. To achieve this goal at the lowest cost to society, the programme employs both regulatory and market based approaches for controlling air pollution.

CHAPTER 13

# Water Pollution

Water pollution is the contamination of water bodies (e.g. lakes, rivers, oceans and groundwater).

Water pollution affects plants and organisms living in these bodies of water; and, in almost all cases the effect is damaging not only to individual species and populations, but also to the natural biological communities.

Water pollution occurs when pollutants are discharged directly or indirectly into water bodies without adequate treatment to remove harmful compounds.

Water pollution is a major problem in the global context. It has been suggested that it is the leading worldwide cause of deaths and diseases, and that it accounts for the deaths of more than 14,000 people daily. An estimated 700 million Indians have no access to a proper toilet, and 1,000 Indian children die of diarrhoeal sickness every day. Some 90 per cent of China's cities suffer from some degree of water pollution, and nearly 500 million people lack access to safe drinking water. In addition to the acute problems of water pollution in developing countries, industrialized countries continue to struggle with pollution problems as well. In the most recent national report on water quality in the United States, 45 per cent of assessed stream miles, 47 per cent of assessed lake acres, and 32 per cent of assessed bay and estuarine square miles were classified as polluted.

Water is typically referred to as polluted when it is impaired by anthropogenic contaminants and either does not support a human use, like serving as drinking water, and/or undergoes a marked shift in its ability to support its constituent biotic communities, such as fish. Natural phenomena such as volcanoes, algae blooms, storms, and earthquakes also cause major changes in water quality and the ecological status of water.

### Water Pollution Categories

Surface water and groundwater have often been studied and managed as separate resources, although they are interrelated. Sources of surface water pollution are generally grouped into two categories based on their origin.

### Point Source Pollution

Point source pollution refers to contaminants that enter a waterway through a discrete conveyance, such as a pipe or ditch. Examples of sources in this category include discharges from a sewage treatment plant, a factory, or a city storm drain. The U.S. Clean Water Act (CWA) defines point source for regulatory enforcement purposes. The CWA definition of point source was amended in 1987 to include municipal storm sewer systems, as well as industrial stormwater, such as from construction sites.

### Non-point Source Pollution

Non-point source (NPS) pollution refers to diffuse contamination that does not originate from a single discrete source. NPS pollution is often the cumulative effect of small amounts of contaminants gathered from a large area. The leaching out of nitrogen compounds from agricultural land which has been fertilized is a typical example. Nutrient runoff in stormwater from 'sheet flow' over an agricultural field or a forest are also cited as examples of NPS pollution.

Contaminated storm water washed off of parking lots, roads and highways, called urban runoff, is sometimes included under the category of NPS pollution. However, this runoff is typically channeled into storm drain systems and discharged through pipes to local surface waters, and is a point source. However where such water is not channelled and drains directly to ground it is a non-point source.

Interactions between groundwater and surface water are complex. Consequently, groundwater pollution, sometimes referred to as groundwater contamination, is not as easily classified as surface water pollution. By its very nature, groundwater aquifers are susceptible to contamination from sources that may not directly affect surface water bodies, and the distinction of point *vs*. non-point source may be irrelevant. A spill or ongoing releases of chemical or radionuclide contaminants into soil (located away from a surface water body) may not create point source or non-point source pollution, but can contaminate the aquifer below, defined as a toxin plume. The movement of the plume, a plume front, can be part of a Hydrological transport model or Groundwater model. Analysis of groundwater contamination may focus on the soil characteristics and site geology, hydrogeology, hydrology, and the nature of the contaminants.

## Causes of Water Pollution

The specific contaminants leading to pollution in water include a wide spectrum of chemicals, pathogens, and physical or sensory changes such as elevated temperature and discoloration. While many of the chemicals and substances that are regulated may be naturally occurring (calcium, sodium, iron, manganese, etc.) the concentration is often the key in determining what is a natural component of water, and what is a contaminant.

Oxygen-depleting substances may be natural materials, such as plant matter (e.g. leaves and grass) as well as man-made chemicals. Other natural and anthropogenic substances may cause turbidity (cloudiness) which blocks light and disrupts plant growth, and clogs the gills of some fish species.

Many of the chemical substances are toxic. Pathogens can produce waterborne diseases in either human or animal hosts. Alteration of water's physical chemistry includes acidity (change in pH), electrical conductivity, temperature, and eutrophication. Eutrophication is an increase in the concentration of chemical nutrients in an ecosystem to an extent that increases in the primary productivity of the ecosystem. Depending on the degree of eutrophication, subsequent negative environmental effects such as anoxia (oxygen depletion) and severe reductions in water quality may occur, affecting fish and other animal populations.

## Pathogens

Coliform bacteria are a commonly used bacterial indicator of water pollution, although not an actual cause of disease. Other microorganisms sometimes found in surface waters which have caused human health problems include:

- Burkholderia pseudomallei
- Cryptosporidium parvum
- Giardia lamblia
- Salmonella
- Novovirus and other viruses
- Parasitic worms (helminths).

High levels of pathogens may result from inadequately treated sewage discharges. This can be caused by a sewage plant designed with less than secondary treatment (more typical in less-developed countries). In developed countries, older cities with aging infrastructure may have leaky sewage collection systems (pipes, pumps, valves), which can cause sanitary sewer overflows. Some cities also have combined sewers, which may discharge untreated sewage during rain storms.

Pathogen discharges may also be caused by poorly managed livestock operations.

**Chemical and other Contaminants**

Contaminants may include organic and inorganic substances.

Organic water pollutants include:

- Detergents
- Disinfection by-products found in chemically disinfected drinking water, such as chloroform
- Food processing waste, which can include oxygen-demanding substances, fats and grease
- Insecticides and herbicides, a huge range of organohalides and other chemical compounds
- Petroleum hydrocarbons, including fuels ( gasoline, diesel fuel, jet fuels, and fuel oil) and lubricants (motor oil), and fuel combustion byproducts, from stormwater runoff
- Tree and bush debris from logging operations
- solvents, from improper storage. Chlorinated solvents, which are dense non-aqueous phase liquids (DNAPLs), may fall to the bottom of reservoirs, since they don't mix well with water and are denser.
- Various chemical compounds found in personal hygiene and cosmetic products.

Inorganic water pollutants include:

- Acidity caused by industrial discharges (especially sulfur dioxide from power plants)
- Ammonia from food processing waste
- Chemical waste as industrial by-products
- Fertilizers containing nutrients—nitrates and phosphates–which are found in stormwater runoff from agriculture, as well as commercial and residential use.
- Heavy metals from motor vehicles (via urban stormwater runoff) and acid mine drainage.

- Silt [sediment) in runoff from construction sites, logging, slash and burn practices or land clearing sites.

Macroscopic pollution—large visible items polluting the water—may be termed 'floatables' in an urban stormwater context, or marine debris when found on the open seas, and can include such items as:

- Trash (e.g. paper, plastic, or food waste) discarded by people on the ground, and that are washed by rainfall into storm drains and eventually discharged into surface waters.
- Nurdles, small ubiquitous waterborne plastic pellets.
- Shipwrecks, large derelict ships.

## Thermal Pollution

Thermal pollution is the rise or fall in the temperature of a natural body of water caused by human influence. A common cause of thermal pollution is the use of water as a coolant by power plants and industrial manufacturers. Elevated water temperatures decreases oxygen levels (which can kill fish) and affects ecosystem composition, such as invasion by new thermophilic species. Urban runoff may also elevate temperature in surface waters.

Thermal pollution can also be caused by the release of very cold water from the base of reservoirs into warmer rivers.

Most water pollutants are eventually carried by rivers into the oceans. In some areas of the world the influence can be traced hundred miles from the mouth by studies using hydrology transport models. Advanced computer models such as SWMM or the DSSAM Model have been used in many locations worldwide to examine the fate of pollutants in aquatic systems. Indicator filter feeding species such as copepods have also been used to study pollutant fates in the New York Bight, for example. The highest toxin loads are not directly at the mouth of the Hudson River, but 100 kilometres south, since several days are required for incorporation into planktonic tissue. The Hudson discharge flows south along the coast due to coriolis force. Further south then are areas of oxygen depletion, caused by chemicals using up oxygen and by algae blooms, caused by excess nutrients from algal cell death and decomposition. Fish and shellfish kills have been reported, because toxins climb the food chain after small fish consume copepods, then large fish eat smaller fish, etc. Each successive step up the food chain causes a stepwise concentration of pollutants such as heavy metals (e.g. mercury) and persistent organic pollutants such as DDT. This is known as biomagnification, which is occasionally used interchangeably with bioaccumulation.

Large gyres (vortexes) in the oceans trap floating plastic debris. The North Pacific Gyre for example has collected the so-called 'Great Pacific Garbage Patch' that is now estimated at 100 times the size of Texas. Many of these long-lasting pieces wind up in the stomachs of marine birds and animals. This results in obstruction of digestive pathways which leads to reduced appetite or even starvation.

Many chemicals undergo reactive decay or chemically change especially over long periods of time in groundwater reservoirs. A noteworthy class of such chemicals is the chlorinated hydrocarbons such as trichloroethylene (used in industrial metal degreasing and electronics manufacturing) and tetrachloroethylene used in the dry cleaning industry (note latest advances in liquid carbon dioxide in dry cleaning that avoids all use of chemicals). Both of these chemicals, which are carcinogens themselves, undergo partial decomposition reactions, leading to new hazardous chemicals (including dichloroethylene and vinyl chloride).

Groundwater pollution is much more difficult to abate than surface pollution because groundwater can move great distances through unseen aquifers. Non-porous aquifers such as clays partially purify water of bacteria by simple filtration (adsorption and absorption), dilution, and, in some cases, chemical reactions and biological activity: however, in some cases, the pollutants merely transform to soil contaminants. Groundwater that moves through cracks and caverns is not filtered and can be transported as easily as surface water. In fact, this can be aggravated by the human tendency to use natural sinkholes as dumps in areas of Karst topography.

Water pollution may be analyzed through several broad categories of methods: physical, chemical and biological. Most involve collection of samples, followed by specialized analytical tests. Some methods may be conducted *in situ*, without sampling, such as temperature. Government agencies and research organizations have published standardized, validated analytical test methods to facilitate the comparability of results from disparate testing events.

## Sampling

Sampling of water for physical or chemical testing can be done by several methods, depending on the accuracy needed and the characteristics of the contaminant. Many contamination events are sharply restricted in time, most commonly in association with rain events. For this reason 'grab' samples are often inadequate for fully quantifying contaminant levels. Scientists gathering this type of data often employ auto-sampler devices that pump increments of water at either time or discharge intervals.

Sampling for biological testing involves collection of plants and/or animals from the surface water body. Depending on the type of assessment, the organisms may be identified for biosurveys (population counts) and returned to the water body, or they may be dissected for bioassays to determine toxicity.

## Physical Testing

Common physical tests of water include temperature, solids concentration like total suspended solids (TSS) and turbidity.

## Chemical Testing

Water samples may be examined using the principles of analytical chemistry. Many published test methods are available for both organic and inorganic compounds. Frequently used methods include pH, biochemical oxygen demand (BOD), chemical oxygen demand (COD), nutrients (nitrate and phosphorus compounds), metals (including copper, zinc, cadmium, lead and mercury), oil and grease, total petroleum hydrocarbons (TPH), and pesticides.

## Domestic Sewage

Domestic sewage is 99.9 per cent pure water, the other 0.1 per cent are pollutants. While found in low concentrations, these pollutants pose risk on a large scale. In urban areas, domestic sewage is typically treated by centralized sewage treatment plants. In the U.S., most of these plants are operated by local government agencies, frequently referred to as publicly owned treatment works (POTW). Municipal treatment plants are designed to control conventional pollutants: BOD and suspended solids. Well-designed and operated systems (i.e., secondary treatment or better) can remove 90 per cent or more of these pollutants. Some plants have additional sub-systems to treat nutrients and pathogens. Most municipal plants are not designed to treat toxic pollutants found in industrial wastewater.

Cities with sanitary sewer overflows or combined sewer overflows employ one or more engineering approaches to reduce discharges of untreated sewage, including:

- utilizing a green infrastructure approach to improve stormwater management capacity throughout the system, and reduce the hydraulic overloading of the treatment plant;
- repair and replacement of leaking and malfunctioning equipment;
- increasing overall hydraulic capacity of the sewage collection system (often a very expensive option).

A household or business not served by a municipal treatment plant may have an individual septic tank, which treats the wastewater on site and discharges into the soil. Alternatively, domestic wastewater may be sent to a nearby privately owned treatment system (e.g. in a rural community).

### Industrial Wastewater

Some industrial facilities generate ordinary domestic sewage that can be treated by municipal facilities. Industries that generate wastewater with high concentrations of conventional pollutants (e.g. oil and grease), toxic pollutants (e.g. heavy metals, volatile organic compounds) or other non-conventional pollutants such as ammonia, need specialized treatment systems. Some of these facilities can install a pre-treatment system to remove the toxic components, and then send the partially treated wastewater to the municipal system. Industries generating large volumes of wastewater typically operate their own complete on-site treatment systems.

Some industries have been successful at redesigning their manufacturing processes to reduce or eliminate pollutants, through a process called pollution prevention.

Heated water generated by power plants or manufacturing plants may be controlled with:

- cooling ponds, man-made bodies of water designed for cooling by evaporation, convection, and radiation;
- cooling towers, which transfer waste heat to the atmosphere through evaporation and/or heat transfer;
- cogeneration, a process where waste heat is recycled for domestic and/or industrial heating purposes.

Non-point source controls Sediment (loose soil) washed off fields is the largest source of agricultural pollution in the United States. Farmers may utilize erosion controls to reduce runoff flows and retain soil on their fields. Common techniques include contour plowing, crop mulching, crop rotation, planting perennial crops and installing riparian buffers.

Nutrients (nitrogen and phosphorus) are typically applied to farmland as commercial fertilizer; animal manure; or spraying of municipal or industrial wastewater (effluent) or sludge. Nutrients may also enter runoff from crop residues, irrigation water, wildlife, and atmospheric deposition. Farmers can develop and implement nutrient management plans to reduce excess application of nutrients.

To minimize pesticide impacts, farmers may use Integrated Pest Management (IPM) techniques (which can include biological pest control) to maintain control over pests, reduce reliance on chemical pesticides, and protect water quality.

Farms with large livestock and poultry operations, such as factory farms, are called *concentrated animal feeding operations* or *confined animal feeding operations* in the U.S. and are being subject to increasing government regulation. Animal slurries are usually treated by containment in lagoons before disposal by spray or trickle application to grassland. Constructed wetlands are sometimes used to facilitate treatment of animal wastes, as are anaerobic lagoons. Some animal slurries are treated by mixing with straw and composted at high temperature to produce a bacteriologically sterile and friable manure for soil improvement.

Effective control of urban runoff involves reducing the velocity and flow of stormwater, as well as reducing pollutant discharges. Local governments use a variety of stormwater management techniques to reduce the effects of urban runoff. These techniques, called best management practices (BMPs) in the U.S., may focus on water quantity control, while others focus on improving water quality, and some perform both functions.

Pollution prevention practices include low impact development techniques, installation of green roofs and improved chemical handling (e.g. management of motor fuels & oil, fertilizers and pesticides). Runoff mitigation systems include infiltration basins, bioretention systems, constructed wetlands, retention basins and similar devices.

Thermal pollution from runoff can be controlled by stormwater management facilities that absorb the runoff or direct it into groundwater, such as bioretention systems and infiltration basins. Retention basins tend to be less effective at reducing temperature, as the water may be heated by the sun before being discharged to a receiving stream.

## Eutrophication

Eutrophication *eutrophia* healthy, adequate nutrition, development (Greek) *eutrophe* (German) is a scientific term describing the overfertilization of lakes with nutrients and the changes that occur as a result i.e. the primary productivity of the waterbody. In other terms, it is the 'bloom' or great increase of phytoplankton in a water body. Negative environmental effects include anoxia, or loss of oxygen in the water with severe reductions in fish and other animal populations. Other species (such as Nomura Jellyfish in Japanese waters) may experience an increase in population that negatively affects other species in the local ecosystem.

Eutrophication can be human-caused or natural. Untreated sewage effluent and agricultural run-off carrying fertilizers are examples of human-caused eutrophication. However, it also occurs naturally in situations where nutrients accumulate (e.g. depositional environments), or where they flow into systems on an ephemeral basis. Eutrophication generally promotes excessive plant growth and decay, favouring simple algae and plankton over other more complicated plants, and causes a severe reduction in water quality. Enhanced growth of aquatic vegetation or phytoplankton and algal blooms disrupts normal functioning of the ecosystem, causing a variety of problems such as a lack of oxygen needed for fish and shellfish to survive. The water becomes cloudy, typically coloured a shade of green, yellow, brown, or red. Eutrophication also decreases the value of rivers, lakes, and estuaries for recreation, fishing, hunting, and aesthetic enjoyment. Health problems can occur where eutrophic conditions interfere with drinking water treatment.

Eutrophication was recognized as a pollution problem in European and North American lakes and reservoirs in the mid-20th century. Since then, it has become more widespread. Surveys showed that 54 per cent of lakes in Asia are eutrophic; in Europe, 53 per cent; in North America, 48 per cent; in South America, 41 per cent; and in Africa, 28 per cent.

Although eutrophication is commonly caused by human activities, it can also be a natural process particularly in lakes. Eutrophy occurs in many lakes in temperate grasslands, for instance. Paleolimnologists now recognise that climate change, geology, and other external influences are critical in regulating the natural productivity of lakes. Some lakes also demonstrate the reverse process (meiotrophication), becoming less nutrient rich with time.

Eutrophication can also be a natural process in seasonally inundated tropical floodplains. In the Barotse Floodplain of the Zambezi River, the first floodwaters of the rainy season are usually hypoxic because of material such as cattle manure and previous decay of vegetation which grew during the dry season. These so-called 'red waters' kill many fish. The process can be made worse by the use of fertilizers in crops such as maize, rice, and sugarcane grown on the floodplain.

Human activities can accelerate the rate at which nutrients enter ecosystems. Runoff from agriculture and development, pollution from septic systems and sewers, and other human-related activities increase the flow of both inorganic nutrients and organic substances into ecosystems. Elevated levels of atmospheric compounds of nitrogen can increase nitrogen availability. Phosphorus is often regarded as the main culprit in cases of eutrophication in lakes subjected to 'point source' pollution from sewage pipes. The concentration of algae and the trophic state of lakes correspond well to phosphorus levels in water. Studies conducted in the Experimental Lakes Area in Ontario have shown a relationship between the addition of phosphorus and the rate of eutrophication. Humankind has increased the rate of phosphorus cycling on Earth by four times, mainly due to agricultural fertilizer production and application. Between 1950 and 1995, an estimated 600,000,000 tonnes of phosphorus were applied to Earth's surface, primarily on croplands. Policy changes to control point sources of phosphorus have resulted in rapid control of eutrophication.

### Ocean Waters

Eutrophication is a common phenomenon in coastal waters. In contrast to freshwater systems, nitrogen is more commonly the key limiting nutrient of marine waters; thus, nitrogen levels have greater importance to understanding eutrophication problems in salt water. Estuaries tend to be naturally eutrophic because land-derived nutrients are concentrated where run-off enters a confined channel. Upwelling in coastal systems also promotes increased productivity by conveying deep, nutrient-rich waters to the surface, where the nutrients can be assimilated by algae.

The World Resources Institute has identified 375 hypoxic coastal zones in the world, concentrated in coastal areas in Western Europe, the Eastern and Southern coasts of the US, and East Asia, particularly Japan.

In addition to runoff from land, atmospheric fixed nitrogen can enter the open ocean. A study in 2008 found that this could account for around one third of the ocean's external (non-recycled) nitrogen supply, and up to three per cent of the annual new marine biological production. It has been suggested that accumulating reactive nitrogen in the environment may prove as serious as putting carbon dioxide in the atmosphere.

### Terrestrial Ecosystems

Terrestrial ecosystems are subject to similarly adverse impacts from eutrophication. Increased nitrates in soil are frequently undesirable for plants. Many terrestrial plant species are endangered as a result of soil eutrophication, such as the majority of orchid species in Europe. Meadows, forests, and bogs are characterized by low nutrient content and slowly growing species adapted to those levels, so they can be overgrown by faster growing and more competitive species. In meadows, tall grasses that can take advantage of higher nitrogen levels may change the area so that natural species may be lost. Species-rich fens can be overtaken by reed or reedgrass species. Forest undergrowth affected by run-off from a nearby fertilized field can be turned into a nettle and bramble thicket.

Chemical forms of nitrogen are most often of concern with regard to eutrophication, because plants have high nitrogen requirements so that additions of nitrogen compounds will stimulate plant growth. Nitrogen is not readily available in soil because $N_2$, a gaseous form of nitrogen, is very stable and unavailable directly to higher plants. Terrestrial ecosystems rely on microbial nitrogen fixation to convert $N_2$ into other forms such as nitrates. However, there is a limit to how much nitrogen can be utilized. Ecosystems receiving more nitrogen than the plants require are called nitrogen-saturated. Saturated terrestrial ecosystems then can contribute both inorganic and organic nitrogen to freshwater, coastal, and marine eutrophication, where nitrogen is also typically a limiting nutrient. This is also the case with increased levels of phosphorus. However, because phosphorus is generally much less soluble than nitrogen, it is leached from the soil at a much slower rate than nitrogen. Consequently, phosphorus is much more important as a limiting nutrient in aquatic systems.

When an ecosystem experiences an increase in nutrients, primary producers reap the benefits first. In aquatic ecosystems, species such as algae experience a population increase (called an algal bloom). Algal blooms limit the sunlight available to bottom-dwelling organisms and cause wide swings in the amount of dissolved oxygen in the water. Oxygen is required by all respiring plants and animals and it is replenished in daylight by photosynthesizing plants and algae. Under eutrophic conditions, dissolved oxygen greatly increases during the day, but is greatly reduced after dark by the respiring algae and by microorganisms that feed on the increasing mass of dead algae. When dissolved oxygen levels decline to hypoxic levels, fish and other marine animals suffocate. As a result, creatures such as fish, shrimp, and especially immobile bottom dwellers die off. In extreme cases, anaerobic conditions ensue, promoting growth of bacteria such as *Clostridium botulinum* that produces toxins deadly to birds and mammals. Zones where this occurs are known as dead zones.

### New Species Invasion

Eutrophication may cause competitive release by making abundant a normally limiting nutrient. This process causes shifts in the species composition of ecosystems. For instance, an increase in nitrogen might allow new, competitive species to invade and out-compete original inhabitant species. This has been shown to occur in New England salt marshes.

### Toxicity

Some algal blooms, otherwise called 'nuisance algae' or 'harmful algal blooms', are toxic to plants and animals. Toxic compounds they produce can make their way up the food chain, resulting in animal mortality. Freshwater algal blooms can pose a threat to livestock. When the algae die or are eaten, neuro- and hepatotoxins are released which can kill animals and may pose a threat to humans. An example of algal toxins working their way into humans is the case of shellfish poisoning. Biotoxins created during algal blooms are taken up by shellfish (mussels, oysters), leading to these human foods acquiring the toxicity and poisoning humans. Examples include paralytic, neurotoxic, and diarrhoetic shellfish poisoning. Other marine animals can be vectors for such toxins, as in the case of ciguatera, where it is typically a predator fish that accumulates the toxin and then poisons humans.

In order to gauge how to best prevent eutrophication from occurring, specific sources that contribute to nutrient loading must be identified. There are two common sources of nutrients and organic matter: point and non-point sources.

**Point Sources**

Point sources are directly attributable to one influence. In point sources the nutrient waste travels directly from source to water. Point sources are relatively easy to regulate.

**Non-point Sources**

Non-point source pollution (also known as 'diffuse' or 'runoff' pollution) is that which comes from ill-defined and diffuse sources. Non-point sources are difficult to regulate and usually vary spatially and temporally (with season, precipitation, and other irregular events).

It has been shown that nitrogen transport is correlated with various indices of human activity in watersheds, including the amount of development. Ploughing in Agriculture and development are activities that contribute most to nutrient loading. There are three reasons that non-point sources are especially troublesome.

**Soil Retention**

Nutrients from human activities tend to accumulate in soils and remain there for years. It has been shown that the amount of phosphorus lost to surface waters increases linearly with the amount of phosphorus in the soil. Thus much of the nutrient loading in soil eventually makes its way to water. Nitrogen, similarly, has a turnover time of decades or more.

**Runoff to Surface Water and Leaching to Groundwater**

Nutrients from human activities tend to travel from land to either surface or ground water. Nitrogen in particular is removed through storm drains, sewage pipes, and other forms of surface runoff. Nutrient losses in runoff and leachate are often associated with agriculture. Modern agriculture often involves the application of nutrients onto fields in order to maximise production. However, farmers frequently apply more nutrients than are taken up by crops or pastures. Regulations aimed at minimising nutrient exports from agriculture are typically far less stringent than those placed on sewage treatment plants and other point source polluters.

## Atmospheric Deposition

Nitrogen is released into the air because of ammonia volatilization and nitrous oxide production. The combustion of fossil fuels is a large human-initiated contributor to atmospheric nitrogen pollution. Atmospheric deposition (e.g., in the form of acid rain) can also affect nutrient concentration in water, especially in highly industrialized regions.

## Other Causes

Any factor that causes increased nutrient concentrations can potentially lead to eutrophication. In modeling eutrophication, the rate of water renewal plays a critical role; stagnant water is allowed to collect more nutrients than bodies with replenished water supplies. It has also been shown that the drying of wetlands causes an increase in nutrient concentration and subsequent eutrophication blooms.

## Prevention and Reversal

Eutrophication poses a problem not only to ecosystems, but to humans as well. Reducing eutrophication should be a key concern when considering future policy, and a sustainable solution for everyone, including farmers and ranchers, seems feasible. While eutrophication does pose problems, humans should be aware that natural runoff (which causes algal blooms in the wild) is common in ecosystems and should thus not reverse nutrient concentrations beyond normal levels.

## Effectiveness

Cleanup measures have been mostly, but not completely, successful. Finnish phosphorus removal measures started in the mid-1970s and have targeted rivers and lakes polluted by industrial and municipal discharges. These efforts have had a 90 per cent removal efficiency. Still, some targeted point sources did not show a decrease in runoff despite reduction efforts.

## Minimizing Non-point Pollution: Future Work

Non-point pollution is the most difficult source of nutrients to manage. The literature suggests, though, that when these sources are controlled, eutrophication decreases. The following steps are recommended to minimize the amount of pollution that can enter aquatic ecosystems from ambiguous sources.

## Riparian Buffer Zones

Studies show that intercepting non-point pollution between the source and the water is a successful means of prevention. Riparian buffer zones are interfaces between a flowing body of water and land, and have been created near waterways in an attempt to filter pollutants; sediment and nutrients are deposited here instead of in water. Creating buffer zones near farms and roads is another possible way to prevent nutrients from traveling too far. Still, studies have shown that the effects of atmospheric nitrogen pollution can reach far past the buffer zone. This suggests that the most effective means of prevention is from the primary source.

## Prevention Policy

Laws regulating the discharge and treatment of sewage have led to dramatic nutrient reductions to surrounding ecosystems, but it is generally agreed that a policy regulating agricultural use of fertilizer and animal waste must be imposed. In Japan the amount of nitrogen produced by livestock is adequate to serve the fertilizer needs for the agriculture industry. Thus, it is not unreasonable to command livestock owners to clean up animal waste–which when left stagnant will leach into ground water.

Policy concerning the prevention and reduction of eutrophication can be broken down into four sectors: Technologies, public participation, economic instruments, and cooperation. The term technology is used loosely, referring to a more widespread use of existing methods rather than an appropriation of new technologies. As mentioned before, nonpoint sources of pollution are the primary contributors to eutrophication, and their effects can be easily minimized through common agricultural practices. Reducing the amount of pollutants that reach a watershed can be achieved through the protection of its forest cover, reducing the amount of erosion leeching into a watershed. Also, through the efficient, controlled use of land using sustainable agricultural practices to minimize land degradation, the amount of soil runoff and nitrogen-based fertilizers reaching a watershed can be reduced. Waste disposal technology constitutes another factor in eutrophication prevention. Because a major contributor to the non-point source nutrient loading of water bodies is untreated domestic sewage, it is necessary to provide treatment facilities to highly urbanized areas, particularly those in underdeveloped nations, in which treatment of domestic waste water is a scarcity. The technology to safely and efficiently reuse waste water, both from domestic and industrial sources, should be a primary concern for policy regarding eutrophication.

The role of the public is a major factor for the effective prevention of eutrophication. In order for a policy to have any effect, the public must be aware of their contribution to the problem, and ways in which they can reduce their effects. Programmes instituted to promote participation in the recycling and elimination of wastes, as well as education on the issue of rational water use are necessary to protect water quality within urbanized areas and adjacent water bodies.

Economic instruments, "which include, among others, property rights, water markets, fiscal and financial instruments, charge systems and liability systems, are gradually becoming a substantive component of the management tool set used for pollution control and water allocation decisions." Incentives for those who practice clean, renewable, water management technologies are an effective means of encouraging pollution prevention. By internalizing the costs associated with the negative effects on the environment, governments are able to encourage a cleaner water management.

Because a body of water can have an effect on a range of people reaching far beyond that of the watershed, cooperation between different organizations is necessary to prevent the intrusion of contaminants that can lead to eutrophication. Agencies ranging from state governments to those of water resource management and non-governmental organizations, going as low as the local population, are responsible for preventing eutrophication of water bodies.

### Nitrogen Testing and Modelling

Soil Nitrogen Testing (N-Testing) is a technique that helps farmers optimize the amount of fertilizer applied to crops. By testing fields with this method, farmers saw a decrease in fertilizer application costs, a decrease in nitrogen lost to surrounding sources, or both. By testing the soil and modeling the bare minimum amount of fertilizer needed, farmers reap economic benefits while the environment remains clean and pure.

CHAPTER 14

# Soil Contamination

Soil contamination (soil pollution) is caused by the presence of xenobiotic (human-made) chemicals or other alteration in the natural soil environment. This type of contamination typically arises from the rupture of underground storage tanks, application of pesticides, percolation of contaminated surface water to subsurface strata, oil and fuel dumping, leaching of wastes from landfills or direct discharge of industrial wastes to the soil. The most common chemicals involved are petroleum hydrocarbons, solvents, pesticides, lead and other heavy metals. This occurrence of this phenomenon is correlated with the degree of industrialization and intensities of chemical usage.

The concern over soil contamination stems primarily from health risks, from direct contact with the contaminated soil, vapors from the contaminants, and from secondary contamination of water supplies within and underlying the soil. Mapping of contaminated soil sites and the resulting cleanup are time consuming and expensive tasks, requiring extensive amounts of geology, hydrology, chemistry and computer modeling skills.

It is in North America and Western Europe that the extent of contaminated land is most well known, with many of countries in these areas having a legal framework to identify and deal with this environmental problem; this however may well be just the tip of the iceberg with developing countries very likely to be the next generation of new soil contamination cases.

The immense and sustained growth of the People's Republic of China since the 1970s has exacted a price from the land in increased soil pollution.

The State Environmental Protection Administration believes it to be a threat to the environment, to food safety and to sustainable agriculture. According to a scientific sampling, 150 million mi (100,000 square kilometers) of China's cultivated land have been polluted, with contaminated water being used to irrigate a further 32.5 million mi (21,670 square kilometers) and another 2 million mi (1,300 square kilometers) covered or destroyed by solid waste. In total, the area accounts for one-tenth of China's cultivatable land, and is mostly in economically developed areas. An estimated 12 million tonnes of grain are contaminated by heavy metals every year, causing direct losses of 20 billion yuan (US$2.57 billion).

This type of contamination typically arises from the rupture of underground storage tanks, application of pesticides, percolation of contaminated surface water to subsurface strata, oil and fuel dumping, leaching of wastes from landfills or direct discharge of industrial wastes to the soil. The most common chemicals involved are petroleum hydrocarbons, solvents, pesticides, lead and other heavy metals. This occurrence of this phenomenon is correlated with the degree of industrialization and intensities of chemical usage.

Treated sewage sludge, known in the industry as biosolids, has become controversial as a fertilizer to the land. As it is the byproduct of sewage treatment, it generally contains contaminants such as organisms, pesticides, and heavy metals than other soil.

There is also controversy surrounding the contamination of fertilizers with heavy metals; a series of newspaper articles in the *Seattle Times* made the issue a 'national focus' in the United States, and culminated in a book called *fateful harvest.*

**Health Effects**

Contaminated or polluted soil directly affects human health through direct contact with soil or via inhalation of soil contaminants which have vaporized; potentially greater threats are posed by the infiltration of soil contamination into groundwater aquifers used for human consumption, sometimes in areas apparently far removed from any apparent source of above ground contamination.

Health consequences from exposure to soil contamination vary greatly depending on pollutant type, pathway of attack and vulnerability of the exposed population. Chronic exposure to chromium, lead and other metals, petroleum, solvents, and many pesticide and herbicide formulations can be carcinogenic, can cause congenital disorders, or can cause other chronic health conditions. Industrial or man-made concentrations of naturally-occurring substances, such as nitrate and ammonia associated with livestock manure from agricultural operations, have also been identified as health hazards in soil and groundwater.

Chronic exposure to benzene at sufficient concentrations is known to be associated with higher incidence of leukemia. Mercury and cyclodienes are known to induce higher incidences of kidney damage, some irreversible. PCBs and cyclodienes are linked to liver toxicity. Organophosphates and carbamates can induce a chain of responses leading to neuromuscular blockage. Many chlorinated solvents induce liver changes, kidney changes and depression of the central nervous system. There is an entire spectrum of further health effects such as headache, nausea, fatigue, eye irritation and skin rash for the above cited and other chemicals. At sufficient dosages a large number of soil contaminants can cause death by exposure via direct contact, inhalation or ingestion of contaminants in groundwater contaminated through soil.

## Ecosystem Effects

Not unexpectedly, soil contaminants can have significant deleterious consequences for ecosystems. There are radical soil chemistry changes which can arise from the presence of many hazardous chemicals even at low concentration of the contaminant species. These changes can manifest in the alteration of metabolism of endemic microorganisms and arthropods resident in a given soil environment. The result can be virtual eradication of some of the primary food chain, which in turn have major consequences for predator or consumer species. Even if the chemical effect on lower life forms is small, the lower pyramid levels of the food chain may ingest alien chemicals, which normally become more concentrated for each consuming rung of the food chain. Many of these effects are now well known, such as the concentration of persistent DDT materials for avian consumers, leading to weakening of egg shells, increased chick mortality and potential extinction of species.

Effects occur to agricultural lands which have certain types of soil contamination. Contaminants typically alter plant metabolism, most commonly to reduce crop yields. This has a secondary effect upon soil conservation, since the languishing crops cannot shield the Earth's soil mantle from erosion phenomena. Some of these chemical contaminants have long half-lives and in other cases derivative chemicals are formed from decay of primary soil contaminants.

## Cleanup Options

Cleanup or remediation is analyzed by environmental scientists who utilize field measurement of soil chemicals and also apply computer models for analyzing transport and fate of soil chemicals. There are several principal strategies for remediation:

- Excavate soil and take it to a disposal site away from ready pathways for human or sensitive ecosystem contact. This technique also applies to dredging of bay muds containing toxins.
- Aeration of soils at the contaminated site (with attendant risk of creating air pollution)
- Thermal remediation by introduction of heat to raise subsurface temperatures sufficiently high to volatize chemical contaminants out of the soil for vapour extraction. Technologies include ISTD, electrical resistance heating (ERH), and ET-DSP$^{tm}$.
- Bioremediation, involving microbial digestion of certain organic chemicals. Techniques used in bioremediation include landfarming, biostimulation and bioaugmentating soil biota with commercially available microflora.
- Extraction of groundwater or soil vapour with an active electromechanical system, with subsequent stripping of the contaminants from the extract.
- Containment of the soil contaminants (such as by capping or paving over in place).
- Phytoremediation, or using plants (such as willow) to extract heavy metals.

# Chapter 15

# Light Pollution

Light pollution, also known as photopollution or luminous pollution, is excessive or obtrusive artificial light.

The International Dark-Sky Association (IDA) defines *light pollution* as:

> Any adverse effect of artificial light including sky glow, glare, light trespass, light clutter, decreased visibility at night, and energy waste.

This approach confuses the cause and its result, however. Pollution is the adding-of/added light itself, in analogy to added sound, carbon dioxide, etc. Adverse consequences are multiple; some of them may be not known yet. Scientific definitions thus include the following:

- Alteration of natural light levels in the outdoor environment owing to artificial light sources.
- Light pollution is the alteration of light levels in the outdoor environment (from those present naturally) due to man-made sources of light. Indoor light pollution is such alteration of light levels in the indoor environment due to sources of light, which compromises human health.
- Light pollution is the introduction by humans, directly or indirectly, of artificial light into the environment.

The first two of the above three scientific definitions describe the state of the environment. The third (and newest) one describes the process of polluting by light.

Light pollution obscures the stars in the night sky for city dwellers, interferes with astronomical observatories, and, like any other form of pollution, disrupts ecosystems and has adverse health effects. Light pollution can be divided into two main types:

1. annoying light that intrudes on an otherwise natural or low-light setting; and
2. excessive light (generally indoors) that leads to discomfort and adverse health effects.

Since the early 1980s, a global dark-sky movement has emerged, with concerned people campaigning to reduce the amount of light pollution.

Light pollution is a side effect of industrial civilization. Its sources include building exterior and interior lighting, advertising, commercial properties, offices, factories, streetlights, and illuminated sporting venues. It is most severe in highly industrialized, densely populated areas of North America, Europe, and Japan and in major cities in the Middle East and North Africa like Tehran and Cairo, but even relatively small amounts of light can be noticed and create problems. Like other forms of pollution (such as air, water, and noise pollution) light pollution causes damage to the environment.

Energy conservation advocates contend that light pollution must be addressed by changing the habits of society, so that lighting is used more efficiently, with less waste and less creation of unwanted or unneeded illumination. Several industry groups also recognize light pollution as an important issue. For example, the Institution of Lighting Engineers in the United Kingdom provides its members information about light pollution, the problems it causes, and how to reduce its impact.

Since not everyone is irritated by the same lighting sources, it is common for one person's light "pollution" to be light that is desirable for another. One example of this is found in advertising, when an advertiser wishes for particular lights to be bright and visible, even though others find them annoying. Other types of light pollution are more certain. For instance, light that *accidentally* crosses a property boundary and annoys a neighbor is generally wasted and pollutive light.

Disputes are still common when deciding appropriate action, and differences in opinion over what light is considered reasonable, and who should be responsible, mean that negotiation must sometimes take place between parties. Where objective measurement is desired, light levels can be quantified by field measurement or mathematical modeling, with results typically displayed as an isophote map or light contour map. Authorities have also taken a variety of measures for dealing with light pollution, depending on the interests, beliefs and understandings of the society involved. Measures range from doing nothing at all, to implementing strict laws and regulations about how lights may be installed and used.

**Types**

*Light pollution* is a broad term that refers to multiple problems, all of which are caused by inefficient, unappealing, or (arguably) unnecessary use of artificial light. Specific categories of light pollution include light trespass, over-illumination, glare, light clutter, and skyglow. A single offending light source often falls into more than one of these categories.

Light trespass occurs when unwanted light enters one's property, for instance, by shining over a neighbor's fence. A common light trespass problem occurs when a strong light enters the window of one's home from the outside, causing problems such as sleep deprivation or the blocking of an evening view.

A number of cities in the U.S. have developed standards for outdoor lighting to protect the rights of their citizens against light trespass. To assist them, the International Dark-Sky Association has developed a set of model lighting ordinances. The Dark-Sky Association was started to reduce the light going up into the sky which reduces visibility of stars, see sky glow below. This is any light which is emitted more than 90 degrees above nadir. By limiting light at this 90 degree mark they have also reduced the light output in the 80-90 degree range which creates most of the light trespass issues. U.S. federal agencies may also enforce standards and process complaints within their areas of jurisdiction. For instance, in the case of light trespass by white strobe lighting from communication towers in excess of FAA minimum lighting requirements the Federal Communications Commission maintains an Antenna Structure Registration database information which citizens may use to identify offending structures and provides a mechanism for processing consumer inquiries and complaints. The US Green Building Council (USGBC) has also incorporated into their environment friendly building standard known as LEED, a credit for reducing the amount of light trespass and sky glow.

Light trespass can be reduced by selecting light fixtures which limit the amount of light emitted more than 80 degrees above the nadir. The IESNA definitions include full cutoff (0%), cutoff (10%), and semi-cutoff (20%).

Over-illumination is the excessive use of light. Specifically within the United States, over-illumination is responsible for approximately two million barrels of oil per day in energy wasted. This is based upon U.S. consumption of equivalent of 50 million barrels per day (7,900,000 $m^3/d$) of petroleum. It is further noted in the same U.S. Department of Energy source that over 30 per cent of all energy is consumed by commercial, industrial and residential sectors. Energy audits of existing buildings demonstrate that the lighting component of residential, commercial and industrial uses consumes about 20 to 40 per cent of those land uses, variable with region and land use. (Residential use lighting consumes only 10 to 30 per cent of the energy bill while commercial buildings major use is lighting. Thus lighting energy accounts for about four or five million barrels of oil (equivalent) per day. Again energy audit data demonstrates that about 30 to 60 per cent of energy consumed in lighting is unneeded or gratuitous.

An alternative calculation starts with the fact that commercial building lighting consumes in excess of 81.68 terawatts (1999 data) of electricity, according to the U.S. DOE. Thus commercial lighting alone consumes about four to five million barrels per day (equivalent) of petroleum, in line with the alternate rationale above to estimate U.S. lighting energy consumption.

Over-illumination stems from several factors:

- Not using timers, occupancy sensors or other controls to extinguish lighting when not needed.
- Improper design, especially of workplace spaces, by specifying higher levels of light than needed for a given task.
- Incorrect choice of fixtures or light bulbs, which do not direct light into areas as needed.
- Improper selection of hardware to utilize more energy than needed to accomplish the lighting task.
- Incomplete training of building managers and occupants to use lighting systems efficiently.
- Inadequate lighting maintenance resulting in increased stray light and energy costs.
- 'Daylight lighting' can be required by citizens to reduce crime or by shop owners to attract customers, so over-illumination can be a design choice, not a fault. In both cases target achievement is questionable.
- Substitution of old mercury lamps with more efficient sodium or metal halide lamps using the same electrical power.
- Indirect lighting techniques, such as lighting a vertical wall to bounce photons on the ground.

Most of these issues can be readily corrected with available, inexpensive technology, and with landlord/tenant practices that create barriers to rapid correction of these matters. Most importantly public awareness would need to improve for industrialized countries to realize the large payoff in reducing over-illumination.

## Light Clutter

Light clutter refers to excessive groupings of lights. Groupings of lights may generate confusion, distract from obstacles (including those that they may be intended to illuminate), and potentially cause accidents. Clutter is particularly noticeable on roads where the street lights are badly designed, or where brightly lit advertising surrounds the roadways. Depending on the motives of the person or organization who installed the lights, their placement and design may even be intended to distract drivers, and can contribute to accidents.

Clutter may also present a hazard in the aviation environment if aviation safety lighting must compete for pilot attention with non-relevant lighting. For instance, runway lighting may be confused with an array of suburban commercial lighting and aircraft collision avoidance lights may be confused with ground lights.

## Skyglow

Skyglow refers to the 'glow' effect that can be seen over populated areas. It is the combination of all light reflected from what it has illuminated escaping up into the sky and from *all* of the badly directed light in that area that also escapes into the sky, being scattered (redirected) by the atmosphere back toward the ground. This scattering is very strongly related to the wavelength of the light when the air is very clear (with very little aerosols). Rayleigh scattering dominates in such clear air, making the sky appear blue in the daytime. When there is significant aerosol (typical of most modern polluted conditions), the scattered light has less dependence on wavelength, making a whiter daytime sky. Because of this Rayleigh effect, and because of the eye's increased sensitivity to white or blue-rich light sources when adapted to very low light levels (see Purkinje effect), white or blue-rich light contributes significantly more to sky-glow than an equal amount of yellow light. Sky glow is of particular irritation to astronomers, because it reduces contrast in the night sky to the extent where it may even become impossible to see any but the brightest stars.

The Bortle Dark-Sky Scale, originally published in *Sky & Telescope* magazine, is sometimes used (by groups like the U.S. National Park Service to quantify skyglow and general sky clarity. The nine-class scale rates the darkness of the night sky and the visibility of its phenomena, such as the gegenschein and the zodiacal light (easily masked by skyglow), providing a detailed description of each level on the scale (with Class 1 being the best).

Light is particularly problematic for amateur astronomers, whose ability to observe the night sky from their property is likely to be inhibited by any stray light from nearby. Most major optical astronomical observatories are surrounded by zones of strictly enforced restrictions on light emissions.

'Direct' skyglow can be reduced by selecting lighting fixtures which limit the amount of light emitted more than 90 degrees above the nadir. The IESNA definitions include full cutoff (0%), cutoff (2.5%), and semi-cutoff (5%). 'Indirect' skyglow produced by reflections from vertical and horizontal surfaces is harder to manage; the only effective method for preventing it is by minimizing over-illumination.

Measuring the effect of sky glow on a global scale is a complex procedure. The natural atmosphere is not completely dark, even in the absence of terrestrial sources of light and illumination from the Moon. This is caused by two main sources: *airglow* and *scattered light*.

At high altitudes, primarily above the mesosphere, there is enough UV radiation from the sun of very short wavelength that ionization occurs. When these ions collide with electrically neutral particles they recombine and emit photons in the process, causing airglow. The degree of ionization is sufficiently large to allow a constant emission of radiation even during the night when the upper atmosphere is in the Earth's shadow. Lower in the atmosphere all of the solar photons with energies above the ionization potential of $N_2$ and $O_2$ have already been absorbed by the higher layers and thus no appreciable ionization occurs.

Apart from emitting light, the sky also scatters incoming light, primarily from distant stars and the Milky Way, but also the zodiacal light, sunlight that is reflected and backscattered from interplanetary dust particles.

The amount of airglow and zodiacal light is quite variable (depending, amongst other things on sunspot activity and the Solar cycle) but given optimal conditions the darkest possible sky has a brightness of about 22 magnitude/square arcsecond. If a full moon is present, the sky brightness increases to 18 magnitude/sq. arcsecond, 40 times brighter than the darkest sky. In densely populated areas a sky brightness of 17 magnitude/sq. arcsecond is not uncommon, or as much as 100 times brighter than is natural.

To precisely measure how bright the sky gets, night time satellite imagery of the earth is used as raw input for the number and intensity of light sources. These are put into a physical model of scattering due to air molecules and aerosoles to calculate cumulative sky brightness. Maps that show the enhanced sky brightness have been prepared for the entire world.

Inspection of the area surrounding Madrid reveals that the effects of light pollution caused by a single large conglomeration can be felt up to 100 km (62 mi) away from the center. Global effects of light pollution are also made obvious. The entire area consisting of southern England, Netherlands, Belgium, west Germany, and northern France have a sky brightness of at least two to four times above normal. The only place in continental Europe where the sky can attain its natural darkness is in northern Scandinavia.

In North America the situation is comparable. From the east coast to west Texas up to the Canadian border there is very significant global light pollution.

Lighting is responsible for one-fourth of all electricity consumption worldwide and case studies have shown that several forms of over-illumination constitute energy wastage, including non-beneficial upward direction of night-time lighting. In 2007, Terna, the company responsible for managing electricity flow in Italy, reported a saving of 645.2 million kWh in electricity consumption during the daylight saving period from April to October. It attributes this saving to the delayed need for artificial lighting during the evenings.

## In Australia

Public lighting is the single largest source of local government's greenhouse gas emissions, typically accounting for 30 to 50 per cent of their emissions. There are 1.94 million public lights–one for every 10 Australians—that annually cost A$210 million, use 1,035 GWh of electricity and are responsible for 1.15 million tonnes of CO2 emissions. Current public lighting in Australia, particularly for minor roads and streets, uses large amounts of energy and financial resources, while often failing to provide high quality lighting. There are many ways to improve lighting quality while reducing energy use and greenhouse gas emissions as well as lowering costs.

Medical research on the effects of excessive light on the human body suggests that a variety of adverse health effects may be caused by light pollution or excessive light exposure, and some lighting design textbooks use human health as an explicit criterion for proper interior lighting. Health effects of over-illumination or improper spectral composition of light may include: increased headache incidence, worker fatigue, medically defined stress, decrease in sexual function and increase in anxiety. Likewise, animal models have been studied demonstrating unavoidable light to produce adverse effect on mood and anxiety. For those who need to be awake at night, light at night also has an acute effect on alertness and mood.

Common levels of fluorescent lighting in offices are sufficient to elevate blood pressure by about eight points. Specifically within the USA, there is evidence that levels of light in most office environments lead to increased stress as well as increased worker errors.

Several published studies also suggest a link between exposure to light at night and risk of breast cancer, due to suppression of the normal nocturnal production of melatonin. In 1978 Cohen et al. proposed that reduced production of the hormone melatonin might increase the risk of breast cancer and citing 'environmental lighting' as a possible causal fa^tor. Researchers at the National Cancer Institute (NCI) and National Institute of Environmental Health Sciences have also concluded a study that suggests that artificial light during the night can be a factor for breast cancer.

In 2007, 'shift work that involves circadian disruption' was listed as a probable carcinogen by the World Health Organization's International Agency for Research on Cancer. Multiple studies have documented a correlation between night shift work and the increased incidence of breast cancer.

A good review of current knowledge of the health consequences of exposure to artificial light at night and an explanation of the causal mechanisms has been published in the Journal of Pineal Research in 2007.

A more recent discussion (2009), written by Professor Steven Lockley, Harvard Medical School, can be found in the CfDS handbook 'Blinded by the Light?'. "Human health implications of light pollution" states that "... light intrusion, even if dim, is likely to have measurable effects on sleep disruption and melatonin suppression. Even if these effects are relatively small from night to night, continuous chronic circadian, sleep and hormonal disruption may have longer-term health risks". The New York Academy of Sciences hosted a meeting in 2009 on Circadian Disruption and Cancer. Forty Danish female shift workers in 2009 were awarded compensation for breast cancer "caused" by shift work made possible by light at night - the most common cause of light pollution.

In June 2009, the American Medical Association developed a policy in support of control of light pollution. News about the decision emphasized glare as a public health hazard leading to unsafe driving conditions. Especially in the elderly, glare produces loss of contrast, obscuring night vision.

Light pollution poses a serious threat to wildlife, having negative impacts on plant and animal physiology. Light pollution can confuse animal navigation, alter competitive interactions, change predator-prey relations, and cause physiological harm. The rhythm of life is orchestrated by the natural diurnal patterns of light and dark, so disruption to these patterns impacts the ecological dynamics.

Studies suggest that light pollution around lakes prevents zooplankton, such as Daphnia, from eating surface algae, helping cause algal blooms that can kill off the lakes' plants and lower water quality. Light pollution may also affect ecosystems in other ways. For example, lepidopterists and entomologists have documented that nighttime light may interfere with the ability of moths and other nocturnal insects to navigate. Night-blooming flowers that depend on moths for pollination may be affected by night lighting, as there is no replacement pollinator that would not be affected by the artificial light. This can lead to species decline of plants that are unable to reproduce, and change an area's longterm ecology.

A 2009 study also suggests deleterious impacts on animals and ecosystems because of perturbation of polarized light or artificial polarisation of light (even during the day, because direction of natural polarization of sun light and its reflexion is a source of information for a lot of animals). This form of pollution is named polarized light pollution (PLP). Unnatural polarized light sources can trigger maladaptive behaviours in polarization-sensitive taxa and alter ecological interactions.

Lights on tall structures can disorient migrating birds. Estimates by the U.S. Fish and Wildlife Service of the number of birds killed after being attracted to tall towers range from 4 to 5 million per year to an order of magnitude higher. The Fatal Light Awareness Programme (FLAP) works with building owners in Toronto, Canada and other cities to reduce mortality of birds by turning out lights during migration periods.

Similar disorientation has also been noted for bird species migrating close to offshore production and drilling facilities. Studies carried out by Nederlandse Aardolie Maatschappij b.v. (NAM) and Shell have led to development and trial of new lighting technologies in the North Sea. In early 2007, the lights were installed on the Shell production platform L15. The experiment proved a great success since the number of birds circling the platform declined by 50 to 90 per cent.

Sea turtle hatchlings emerging from nests on beaches are another casualty of light pollution. It is a common misconception that hatchling sea turtles are attracted to the moon. Rather, they find the ocean by moving away from the dark silhouette of dunes and their vegetation, a behaviour with which artificial lights interfere. The breeding activity and reproductive phenology of toads, however, are cued by moonlight. Juvenile seabirds may also be disoriented by lights as they leave their nests and fly out to sea. Amphibians and reptiles are also affected by light pollution. Introduced light sources during normally dark periods can disrupt levels of melatonin production. Melatonin is a hormone that regulates photoperiodic physiology and behaviour. Some species of frogs and salamanders utilize a light-dependent 'compass' to orient their migratory behaviour to breeding sites. Introduced light can also cause developmental irregularities, such as retinal damage, reduced sperm production, and genetic mutation.

In September 2009, the 9th European Dark-Sky Symposium in Armagh, Northern Ireland had a session on the environmental effects of light at night (LAN). It dealt with bats, turtles, the 'hidden' harms of LAN, and many other topics. The environmental effects of LAN were mentioned as early as 1897, in a *Los Angeles Times* article—the text of which can be obtained from Dr. Travis Longcore of the Urban Wildlands Trust, California. The following is an excerpt from that article, called 'Electricity and English songbirds'.

An English journal has become alarmed at the relation of electricity to songbirds, which it maintains is closer than that of cats and fodder crops. How many of us, it asks, foresee that electricity may extirpate the songbird? . . . With the exception of the finches, all the English songbirds may be said to be insectivorous, and their diet consists chiefly of vast numbers of very small insects which they collect from the grass and herbs before the dew is dry. As the electric light is finding its way for street illumination into the country parts of England, these poor winged atoms are slain by thousands at each light every warm summer evening.... The fear is expressed, that when England is lighted from one end to the other with electricity the song birds will die out from the failure of their food supply.

Skyglow reduces the contrast between stars and galaxies in the sky and the sky itself, making it more difficult to detect fainter objects. This is one factor that has caused newer telescopes to be built in increasingly remote areas. Some astronomers use narrow-band 'nebula filters' which only allow specific wavelengths of light commonly seen in nebulae, or broad-band 'light pollution filters' which are designed to reduce (but not eliminate) the effects of light pollution by filtering out spectral lines commonly emitted by sodium- and mercury-vapor lamps, thus enhancing contrast and improving the view of dim objects such as galaxies and nebulae. Unfortunately this affects colour perception, so these filters cannot be used to visually estimate variable star brightness, and no filter can match the effectiveness of a dark sky for visual or photographic purposes. Due to low surface brightness, the visibility of diffuse sky objects such as nebulae and galaxies is affected by light pollution more than are stars. A simple method for estimating the darkness of a location is to look for the milky way.

Light trespass can impact observations when stray light enters the tube of the telescope from off-axis, and is reflected from surfaces other than the telescope's mirrors (if any) so that it eventually reaches the eyepiece, causing a glow across the field of view since it has not been focused. The usual measures to reduce this glare, if reducing the light directly (e.g., by changing one's location or having the light turned off) is not an option, include flocking the telescope tube and accessories to reduce reflection, and putting a light shield (also usable as a dew shield) on the telescope to reduce light entering from angles other than those near the target. In one Italian regional lighting code this effect of stray light is defined as "optical pollution" due to the fact that there is a direct path from the light source to the 'optic' — the observer's eye or telescope.

Reducing light pollution implies many things, such as reducing sky glow, reducing glare, reducing light trespass, and reducing clutter. The method for best reducing light pollution, therefore, depends on exactly what the problem is in any given instance. Possible solutions include:

- Utilizing light sources of minimum intensity necessary to accomplish the light's purpose.
- Turning lights off using a timer or occupancy sensor or manually when not needed.
- Improving lighting fixtures, so that they direct their light more accurately towards where it is needed, and with less side effects.
- Adjusting the *type* of lights used, so that the light waves emitted are those that are less likely to cause severe light pollution problems.
- Evaluating existing lighting plans, and re-designing some or all of the plans depending on whether existing light is actually needed.

The use of *full cutoff* lighting fixtures, as much as possible, is advocated by most campaigners for the reduction of light pollution. It is also commonly recommended that lights be spaced appropriately for maximum efficiency, and that lamps within the fixtures not be overpowered.

Full cutoff fixtures first became available in 1959 with the introduction of General Electric's M100 fixture.

A full cutoff fixture, when correctly installed, reduces the chance for light to escape above the plane of the horizontal. Light released above the horizontal may sometimes be lighting an intended target, but often serves no purpose. When it enters into the atmosphere, light contributes to sky glow. Some governments and organizations are now considering, or have already implemented, full cutoff fixtures in street lamps and stadium lighting.

The use of full cutoff fixtures may help to reduce sky glow by preventing light from escaping unnecessarily. Full cutoff typically reduces the visibility of the lamp and reflector within a luminaire, so the effects of glare may also be reduced. Campaigners also commonly argue that full cutoff fixtures are more efficient than other fixtures, since light that would otherwise have escaped into the atmosphere may instead be directed towards the ground. However, full cutoff fixtures may also trap more light in the fixture than other types of luminaires, corresponding to lower luminaire efficiency.

The use of full cutoff fixtures may allow for lower wattage lamps to be used in the fixtures, producing the same or sometimes a better effect, due to being more carefully controlled. In every lighting system, some sky glow also results from light reflected from the ground. This reflection can be reduced, however, by being careful to use only the lowest wattage necessary for the lamp, and setting spacing between lights appropriately.

A common criticism of full cutoff lighting fixtures is that they are sometimes not as aesthetically pleasing to look at. This is most likely because historically there has not been a large market specifically for full cutoff fixtures, and because people typically like to see the source of illumination. Due to the specificity with their direction of light, full cutoff fixtures sometimes also require expertise to install for maximum effect.

The effectiveness of using full cutoff roadway lights to combat light pollution has also been called into question. According to design investigations, luminaires with full cutoff distributions (as opposed to *cutoff* or *semi cutoff*, compared here have to be closer together to meet the same light level, uniformity and glare requirements specified by the IESNA .These simulations optimized the height and spacing of the lights while constraining the overall design to meet the IESNA requirements, and then compared total uplight and energy consumption of different luminaire designs and powers. Cutoff designs performed better than full cutoff designs, and semi-cutoff performed better than either cutoff or full cutoff. This indicates that, in roadway installations, over-illumination or poor uniformity produced by full cutoff fixtures may be more detrimental than direct uplight created by fewer cutoff or semi-cutoff fixtures. Therefore, the overall performance of existing systems could be improved more by reducing the number of luminaires than by switching to full cutoff designs.

However, using the definition of 'light pollution' from some Italian regional bills (i.e., "every irradiance of artificial light outside competence areas and particularly upward the sky") only full cutoff design prevents light pollution.The Italian Lombardy region, where only full cutoff design is allowed (Lombardy act no. 17/2000, promoted by Cielobuio-coordination for the protection of the night sky), in 2007 had the lowest per capita energy consumption for public lighting in Italy: this information can be verified using data released by Terna company. The same legislation also imposes a minimum distance between street lamps of about four times their height, so full cut off street lamps are the best solution to reduce both light pollution and electrical power usage.

## Adjusting Types of Light Sources

Several different types of light sources exist, each having different properties that affect their appropriateness for certain tasks, particularly efficiency and spectral power distribution. It is often the case that inappropriate light sources have been selected for a task, either due to ignorance or because more sophisticated light sources were unavailable at the time of installation. Therefore, badly chosen light sources often contribute unnecessarily to light pollution and energy waste. By re-assessing and changing the light sources used, it is often possible to reduce energy use and pollutive effects while simultaneously greatly improving efficiency and visibility.

Many astronomers request that nearby communities use low pressure sodium lights as much as possible, because the principal wavelength emitted is comparably easy to work around or in rare cases filter out. The low cost of operating sodium lights is another feature. In 1980, for example, San Jose, California, replaced all street lamps with low pressure sodium lamps, whose light is easier for nearby Lick Observatory to filter out. Similar programmes are now in place in Arizona and Hawaii.

Disadvantages of low pressure sodium lighting are that fixtures must usually be larger than competing fixtures, and that colour cannot be distinguished, due to its emitting principally a single wavelength of light (see security lighting). Due to the substantial size of the lamp, particularly in higher wattages such as 135 W and 180 W, control of light emissions from low pressure sodium luminaires is more difficult. For applications requiring more precise direction of light (such as narrow roadways) the native lamp efficacy advantage of this lamp type is decreased and may be entirely lost compared to high pressure sodium lamps. Allegations that this also leads to higher amounts of light pollution from luminaires running these lamps arise principally because of older luminaires with poor shielding, still widely in use in the UK and in some other locations. Modern low-pressure sodium fixtures with better optics and full shielding, and the decreased skyglow impacts of yellow light preserve the luminous efficacy advantage of low-pressure sodium and result in most cases is less energy consumption and less visible light pollution. Unfortunately, due to continued lack of accurate information, many lighting professionals continue to disparage low-pressure sodium, contributing to its decreased acceptance and specification in lighting standards and therefore its use. Another disadvantage of low-pressure sodium lamps is that some people find the characteristic yellow light very displeasing aesthetically.

Because of the scatter of light by the atmosphere, different sources produce dramatically different amounts of skyglow from the same amount of light sent into the atmosphere.

In some cases, evaluation of existing plans has determined that more efficient lighting plans are possible. For instance, light pollution can be reduced by turning off unneeded outdoor lights, and only lighting stadiums when there are people inside. Timers are especially valuable for this purpose. One of the world's first coordinated *legislative* efforts to reduce the adverse effect of this pollution on the environment began in Flagstaff, Arizona, in the U.S. There, over three decades of ordinance development has taken place, with the full support of the population, often with government support, with community advocates, and with the help of major local observatories, including the United States Naval Observatory Flagstaff Station. Each component helps to educate, protect and enforce the imperatives to intelligently reduce detrimental light pollution.

One example of a lighting plan assessment can be seen in a report originally commissioned by the Office of the Deputy Prime Minister in the United Kingdom, and now available through the Department for Communities and Local Government. The report details a plan to be implemented throughout the UK, for designing lighting schemes in the countryside, with a particular focus on preserving the environment.

In another example, the city of Calgary has recently replaced most residential street lights with models that are comparably energy efficient. The motivation is primarily operation cost and environmental conservation. The costs of installation are expected to be regained through energy savings within six to seven years.

The Swiss Agency for Energy Efficiency (SAFE) uses a concept that promises to be of great use in the diagnosis and design of road lighting, '*consommation électrique spécifique* (*CES*)', which can be translated into English as 'specific electric power consumption (SEC)'. Thus, based on observed lighting levels in a wide range of Swiss towns, SAFE has defined target values for electric power consumption per metre for roads of various categories. Thus, SAFE currently recommends an SEC of 2 to 3 watts per meter for roads of less than 10 metre width (4 to 6 watts per metre for wider roads). Such a measure provides an easily applicable environmental protection constraint on conventional 'norms', which usually are based on the recommendations of lighting manufacturing interests, who may not take into account environmental criteria. In view of ongoing progress in lighting technology, target SEC values will need to be periodically revised downwards.

A newer method for predicting and measuring various aspects of light pollution was described in the journal Lighting Research Technology. Scientists at Rensselaer Polytechnic Institute's Lighting Research Center have developed a comprehensive method called Outdoor Site-Lighting

Performance (OSP), which allows users to quantify, and thus optimize, the performance of existing and planned lighting designs and applications to minimize excessive or obtrusive light leaving the boundaries of a property. OSP can be used by lighting engineers immediately, particularly for the investigation of glow and trespass (glare analyses are more complex to perform and current commercial software does not readily allow them), and can help users compare several lighting design alternatives for the same site.

In the effort to reduce light pollution, researchers have developed a 'Unified System of Photometry', which is a way to measure how much or what kind of street lighting is needed. The Unified System of Photometry allows light fixtures to be designed to reduce energy use while maintaining or improving perceptions of visibility, safety, and security. There was a need to create a new system of light measurement at night because the biological way in which the eye's rods and cones process light is different in nighttime conditions versus daytime conditions. Using this new system of photometry, results from recent studies have indicated that replacing traditional, yellowish, high-pressure sodium (HPS) lights with 'cool' white light sources, such as induction, fluorescent, ceramic metal halide, or LEDs can actually reduce the amount of electric power used for lighting while maintaining or improving visibility in nighttime conditions.

The International Commission on Illumination, also known as the CIE from its French title, le Commission Internationale de l'Eclairage, will soon be releasing its own form of unified photometry for outdoor lighting.

CHAPTER 16

# Noise Pollution

Noise pollution (or environmental noise) is displeasing human, animal or machine-created sound that disrupts the activity or balance of human or animal life. The word *noise* comes from the Latin word *nauseas*, meaning seasickness.

The source of most outdoor noise worldwide is mainly construction and transportation systems, including motor vehicle noise, aircraft noise and rail noise. Poor urban planning may give rise to noise pollution, since side-by-side industrial and residential buildings can result in noise pollution in the residential area.

Indoor and outdoor noise pollution sources include car alarms, emergency service sirens, mechanical equipment, fireworks, compressed air horns, groundskeeping equipment, barking dogs, appliances, lighting hum, audio entertainment systems, electric megaphones, and loud people.

Noise health effects are both health and behavioural in nature. The unwanted sound is called noise. This unwanted sound can damage physiological and psychological health. Noise pollution can cause annoyance and aggression, hypertension, high stress levels, tinnitus, hearing loss, sleep disturbances, and other harmful effects. Furthermore, stress and hypertension are the leading causes to health problems, whereas tinnitus can lead to forgetfulness, severe depression and at times panic attacks.

Chronic exposure to noise may cause noise-induced hearing loss. Older males exposed to significant occupational noise demonstrate significantly reduced hearing sensitivity than their non-exposed peers, though differences in hearing sensitivity decrease with time and the two groups are indistinguishable by age 79. A comparison of Maaban tribesmen, who were insignificantly exposed to transportation or industrial noise, to a typical U.S. population showed that chronic exposure to moderately high levels of environmental noise contributes to hearing loss.

High noise levels can contribute to cardiovascular effects and exposure to moderately high levels during a single eight hour period causes a statistical rise in blood pressure of five to ten points and an increase in stress and vasoconstriction leading to the increased blood pressure noted above as well as to increased incidence of coronary artery disease.

Noise pollution is also a cause of annoyance. A 2005 study by Spanish researchers found that in urban areas households are willing to pay approximately four Euros per decibel per year for noise reduction.

Noise can have a detrimental effect on animals by causing stress, increasing risk of death by changing the delicate balance in predator/prey detection and avoidance, and by interfering with their use of sounds in communication especially in relation to reproduction and in navigation. Acoustic overexposure can lead to temporary or permanent loss of hearing.

An impact of noise on animal life is the reduction of usable habitat that noisy areas may cause, which in the case of endangered species may be part of the path to extinction. Noise pollution has caused the death of certain species of whales that beached themselves after being exposed to the loud sound of military sonar.

Noise also makes species communicate louder, which is called Lombard vocal response. Scientists and researchers have conducted experiments that show whales' song length is longer when submarine-detectors are on. If creatures do not 'speak' loud enough, their voice will be masked by anthropogenic sounds. These unheard voices might be warnings, finding of prey, or preparations of net-bubbling. When one species begins speaking louder, it will mask other species' voice, causing the whole ecosystem to eventually speak louder.

European Robins living in urban environments are more likely to sing at night in places with high levels of noise pollution during the day, suggesting that they sing at night because it is quieter, and their message can propagate through the environment more clearly. The same study showed that daytime noise was a stronger predictor of nocturnal singing than night-time light pollution, to which the phenomenon is often attributed.

Zebra finches become less faithful to their partners when exposed to traffic noise. This could alter a population's evolutionary trajectory by selecting traits, sapping resources normally devoted to other activities and thus lead to profound genetic and evolutionary consequences.

Figures compiled by Rockwool, the mineral wool insulation manufacturer, based on responses from local authorities to a Freedom of Information Act (FOI) request reveal in the period April 2008-2009 UK councils received 315,838 complaints about noise pollution from private residences. This resulted in environmental health officers across the UK serving 8,069 noise abatement notices, or citations under the terms of the Anti-Social Behaviour (Scotland) Act. In the last 12 months, 524 confiscations of equipment have been authorised involving the removal of powerful speakers, stereos and televisions. Westminster City Council has received more complaints per head of population than any other district in the UK with 9,814 grievances about noise, which equates to 42.32 complaints per thousand residents. Eight of the top 10 councils ranked by complaints per 1,000 residents are located in London.

There are a variety of strategies for mitigating roadway noise including: use of noise barriers, limitation of vehicle speeds, alteration of roadway surface texture, limitation of heavy vehicles, use of traffic controls that smooth vehicle flow to reduce braking and acceleration, and tire design. An important factor in applying these strategies is a computer model for roadway noise, that is capable of addressing local topography, meteorology, traffic operations and hypothetical mitigation. Costs of building-in mitigation can be modest, provided these solutions are sought in the planning stage of a roadway project.

Aircraft noise can be reduced to some extent by design of quieter jet engines, which was pursued vigorously in the 1970s and 1980s. This strategy has brought limited but noticeable reduction of urban sound levels. Reconsideration of operations, such as altering flight paths and time of day runway use, has demonstrated benefits for residential populations near airports. FAA sponsored residential retrofit (insulation) programmes initiated in the 1970s has also enjoyed success in reducing interior residential noise in thousands of residences across the United States.

Exposure of workers to Industrial noise has been addressed since the 1930s. Changes include redesign of industrial equipment, shock mounting assemblies and physical barriers in the workplace.

Noise Free America, a national anti-noise pollution organization, regularly lobbies for the enforcement of noise ordinances at all levels of government.

Governments up until the 1970s viewed noise as a "nuisance" rather than an environmental problem. In the United States there are federal standards for highway and aircraft noise; states and local governments typically have very specific statutes on building codes, urban planning and roadway development. In Canada and the EU there are few national, provincial, or state laws that protect against noise.

Noise laws and ordinances vary widely among municipalities and indeed do not even exist in some cities. An ordinance may contain a general prohibition against making noise that is a nuisance, or it may set out specific guidelines for the level of noise allowable at certain times of the day and for certain activities.

Dr. Paul Herman wrote the first comprehensive noise codes in 1975 for Portland, Oregon with funding from the EPA (Environmental Protection Agency) and HUD (Housing and Urban Development). The Portland Noise Code became the basis for most other ordinances for major U.S. and Canadian metropolitan regions.

Most city ordinances prohibit sound above a threshold intensity from trespassing over property line at night, typically between 10 p.m. and 6 a.m., and during the day restricts it to a higher sound level; however, enforcement is uneven. Many municipalities do not follow up on complaints. Even where a municipality has an enforcement office, it may only be willing to issue warnings, since taking offenders to court is expensive.

The notable exception to this rule is the City of Portland Oregon which has instituted an aggressive protection for its citizens with fines reaching as high at $5000 per infraction, with the ability to cite a responsible noise violator multiple times in a single day.

Many conflicts over noise pollution are handled by negotiation between the emitter and the receiver. Escalation procedures vary by country, and may include action in conjunction with local authorities, in particular the police. Noise pollution often persists because only five to ten per cent of people affected by noise will lodge a formal complaint. Many people are not aware of their legal right to quiet and do not know how to register a complaint.

CHAPTER 17

# Radioactive Contamination

Radioactive contamination, also called radiological contamination, is radioactive substances on surfaces, or within solids, liquids or gases (including the human body), where their presence is unintended or undesirable, or the process giving rise to their presence in such places.

Also used less formally to refer to a quantity, namely the activity on a surface (or on a unit area of a surface).

Contamination does not include residual radioactive material remaining at a site after the completion of decommissioning.

The term radioactive contamination may have a connotation that is not intended.

The term radioactive contamination refers only to the presence of radioactivity, and gives no indication of the magnitude of the hazard involved.

The amount of radioactive material released in an accident is called the source term.

Radioactive contamination is typically the result of a spill or accident during the production or use of radionuclides (radioisotopes), an unstable nucleus which has excessive energy. Contamination may occur from radioactive gases, liquids or particles. For example, if a radionuclide used in nuclear medicine is accidentally spilled, the material could be spread by people as they walk around. Radioactive contamination may also be an inevitable result of certain processes, such as the release of radioactive xenon in nuclear fuel reprocessing. In cases that radioactive material cannot be contained, it may be diluted to safe concentrations. Nuclear fallout is the distribution of radioactive contamination by a nuclear explosion. For a discussion of environmental contamination by alpha emitters. Containment is what differentiates radioactive material from radioactive contamination. Therefore, radioactive material in sealed and designated containers is not properly referred to as contamination, although the units of measurement might be the same.

## Radiation Monitoring

The radiation monitoring involves the measurement of radiation dose or radionuclide contamination for reasons related to the assessment or control of exposure to radiation or radioactive substances, and the interpretation of the results. The methodological and technical details of the design and operation of environmental radiation monitoring programmes and systems for different radionuclides, environmental media and types of facility are given in IAEA Safety Standards Series No. RS–G-1.8 and in IAEA Safety Reports Series No. 64.

## Measurement

Radioactive contamination may exist on surfaces or in volumes of material or air. In a nuclear power plant, detection and measurement of radioactivity and contamination is often the job of a Certified Health Physicist.

## Surface Contamination

Surface contamination is usually expressed in units of radioactivity per unit of area. For SI, this is becquerels per square meter (or $Bq/m^2$). Other units such as picoCuries per 100 $cm^2$ or disintegrations per minute per square centimeter (1 $dpm/cm^2$ = 166 2/3 $Bq/m^2$) may be used. Surface contamination may either be fixed or removable. In the case of fixed contamination, the radioactive material cannot by definition be spread, but it is still measurable.

## Hazards

In practice there is no such thing as zero radioactivity. Not only is the entire world constantly bombarded by cosmic rays, but every living creature on earth contains significant quantities of carbon-14 and most (including humans) contain significant quantities of potassium-40. These tiny levels of radiation are not any more harmful than sunlight, but just as excessive quantities of sunlight can be dangerous, so too can excessive levels of radiation.

## Low Level Contamination

The hazards to people and the environment from radioactive contamination depend on the nature of the radioactive contaminant, the level of contamination, and the extent of the spread of contamination. Low levels of radioactive contamination pose little risk, but can still be detected by radiation instrumentation. In the case of low-level contamination by isotopes with a short half-life, the best course of action may be to simply allow the material to naturally decay. Longer-lived isotopes should be cleaned up and properly disposed of, because even a very low level of radiation can be life-threatening when in long exposure to it.

## High Level Contamination

High levels of contamination may pose major risks to people and the environment. People can be exposed to potentially lethal radiation levels, both externally and internally, from the spread of contamination following an accident (or a deliberate initiation) involving large quantities of radioactive material. The biological effects of external exposure to radioactive contamination are generally the same as those from an external radiation source not involving radioactive materials, such as x-ray machines, and are dependent on the absorbed dose.

## Biological Effects

The biological effects of internally deposited radionuclides depend greatly on the activity and the biodistribution and removal rates of the radionuclide, which in turn depends on its chemical form. The biological effects may also depend on the chemical toxicity of the deposited material, independent of its radioactivity. Some radionuclides may be generally distributed throughout the body and rapidly removed, as is the case with tritiated water. Some radionuclides may target specific organs and have much lower removal rates. For instance, the thyroid gland takes up a large percentage of any iodine that enters the body. If large quantities of radioactive iodine are inhaled or ingested, the thyroid may be impaired or destroyed, while other tissues are affected to a lesser extent. Radioactive iodine is a common fission product; it was a major component of the radiation released from the Chernobyl disaster, leading to nine fatal cases of pediatric thyroid cancer and hypothyroidism. On the other hand, radioactive iodine is used in the diagnosis and treatment of many diseases of the thyroid precisely because of the thyroid's selective uptake of iodine.

## Means of Contamination

Radioactive contamination can enter the body through ingestion, inhalation, absorption, or injection. For this reason, it is important to use personal protective equipment when working with radioactive materials. Radioactive contamination may also be ingested as the result of eating contaminated plants and animals or drinking contaminated water or milk from exposed animals. Following a major contamination incident, all potential pathways of internal exposure should be considered.

CHAPTER 18

# Thermal Pollution and Global Warming

Thermal pollution is the degradation of water quality by any process that changes ambient water temperature.

A common cause of thermal pollution is the use of water as a coolant by power plants and industrial manufacturers. When water used as a coolant is returned to the natural environment at a higher temperature, the change in temperature:

*(a)* decreases oxygen supply; and

*(b)* affects ecosystem composition.

Urban runoff—stormwater discharged to surface waters from roads and parking lots—can also be a source of elevated water temperatures.

When a power plant first opens or shuts down for repair or other causes, fish and other organisms adapted to particular temperature range can be killed by the abrupt rise in water temperature known as 'thermal shock'.

Elevated temperature typically decreases the level of dissolved oxygen (DO) in water. The decrease in levels of DO can harm aquatic animals such as fish, amphibians and copepods. Thermal pollution may also increase the metabolic rate of aquatic animals, as enzyme activity, resulting in these organisms consuming more food in a shorter time than if their environment were not changed. An increased metabolic rate may result in fewer resources; the more adapted organisms moving in may have an advantage over organisms that are not used to the warmer temperature. As a result one has the problem of compromising food chains of the old and new environments. Biodiversity can be decreased as a result.

It is known that temperature changes of even one to two degrees Celsius can cause significant changes in organism metabolism and other adverse cellular biology effects. Principal adverse changes can include rendering cell walls less permeable to necessary osmosis, coagulation of cell proteins, and alteration of enzyme metabolism. These cellular level effects can adversely affect mortality and reproduction.

Primary producers are affected by warm water because higher water temperature increases plant growth rates, resulting in a shorter lifespan and species overpopulation. This can cause an algae bloom which reduces oxygen levels.

A large increase in temperature can lead to the denaturing of life-supporting enzymes by breaking down hydrogen- and disulphide bonds within the quaternary structure of the enzymes. Decreased enzyme activity in aquatic organisms can cause problems such as the inability to break down lipids, which leads to malnutrition.

In limited cases, warm water has little deleterious effect and may even lead to improved function of the receiving aquatic ecosystem. This phenomenon is seen especially in seasonal waters and is known as thermal enrichment. An extreme case is derived from the aggregational habits of the manatee, which often uses power plant discharge sites during winter. Projections suggest that manatee populations would decline upon the removal of these discharges.

## Ecological Effects — Cold Water

Releases of unnaturally cold water from reservoirs can dramatically change the fish and macroinvertebrate fauna of rivers, and reduce river productivity. In Australia, where many rivers have warmer temperature regimes, native fish species have been eliminated, and macroinvertebrate fauna have been drastically altered. In the United States, thermal pollution from industrial sources is generated mostly by power plants, petroleum refineries, pulp and paper mills, chemical plants, steel mills and smelters. Heated water from these sources may be controlled with:

- cooling ponds, man-made bodies of water designed for cooling by evaporation, convection, and radiation;
- cooling towers, which transfer waste heat to the atmosphere through evaporation and/or heat transfer;
- cogeneration, a process where waste heat is recycled for domestic and/or industrial heating purposes.

Some facilities use once-through cooling (OTC) systems which do not reduce temperature as effectively as the above systems. For example, the

Potrero Generating Station in San Francisco, which uses OTC, discharges water to San Francisco Bay approximately 10°C (20°F) above the ambient bay temperature. During warm weather, urban runoff can have significant thermal impacts on small streams, as stormwater passes over hot parking lots, roads and sidewalks. Stormwater management facilities that absorb runoff or direct it into groundwater, such as bioretention systems and infiltration basins, can reduce these thermal effects. Retention basins tend to be less effective at reducing temperature, as the water may be heated by the sun before being discharged to a receiving stream.

## Global Warming

Global warming refers to the alleged increase in the average temperature of Earth's near-surface air and oceans since the mid-20th century and its projected continuation. According to the 2007 Fourth Assessment Report by the Intergovernmental Panel on Climate Change (IPCC), global surface temperature increased 0.74 ± 0.18° C (1.33 ± 0.32 °F) during the 20th century. Most of the observed temperature increase since the middle of the 20th century has been caused by increasing concentrations of greenhouse gases, which result from human activity such as the burning of fossil fuel and deforestation. Global dimming, a result of increasing concentrations of atmospheric aerosols that block sunlight from reaching the surface, has partially countered the effects of warming induced by greenhouse gases.

Climate model projections summarized in the latest IPCC report indicate that the global surface temperature is likely to rise a further 1.1 to 6.4° C (2.0 to 11.5° F) during the 21st century. The uncertainty in this estimate arises from the use of models with differing sensitivity to greenhouse gas concentrations and the use of differing estimates of future greenhouse gas emissions. An increase in global temperature will cause sea levels to rise and will change the amount and pattern of precipitation, probably including expansion of subtropical deserts. Warming is expected to be strongest in the Arctic and would be associated with continuing retreat of glaciers, permafrost and sea ice. Other likely effects include changes in the frequency and intensity of extreme weather events, species extinctions, and changes in agricultural yields. Warming and related changes will vary from region to region around the globe, though the nature of these regional variations is uncertain. As a result of contemporary increases in atmospheric carbon dioxide, the oceans have become more acidic, a result that is predicted to continue.

The scientific consensus is that anthropogenic global warming is occurring. Nevertheless, political and public debate continues. The Kyoto

Protocol is aimed at stabilizing greenhouse gas concentration to prevent a 'dangerous anthropogenic interference'. As of November 2009, 187 states had signed and ratified the protocol.

Evidence for warming of the climate system includes observed increases in global average air and ocean temperatures, widespread melting of snow and ice, and rising global average sea level. The most common measure of global warming is the trend in globally averaged temperature near the Earth's surface. Expressed as a linear trend, this temperature rose by 0.74 ± 0.18° C over the period 1906-2005. The rate of warming over the last half of that period was almost double that for the period as a whole (0.13 ± 0.03° C per decade, versus 0.07° C ± 0.02° C per decade). The urban heat island effect is estimated to account for about 0.002° C of warming per decade since 1900. Temperatures in the lower troposphere have increased between 0.13 and 0.22° C (0.22 and 0.4° F) per decade since 1979, according to satellite temperature measurements. Temperature is believed to have been relatively stable over the one or two thousand years before 1850, with regionally varying fluctuations such as the Medieval Warm Period and the Little Ice Age.

Estimates by NASA's Goddard Institute for Space Studies (GISS) and the National Climatic Data Center show that 2005 was the warmest year since reliable, widespread instrumental measurements became available in the late 19th century, exceeding the previous record set in 1998 by a few hundredths of a degree. Estimates prepared by the World Meteorological Organization and the Climatic Research Unit show 2005 as the second warmest year, behind 1998. Temperatures in 1998 were unusually warm because the strongest El Niño in the past century occurred during that year. Global temperature is subject to short-term fluctuations that overlay long term trends and can temporarily mask them. The relative stability in temperature from 2002 to 2009 is consistent with such an episode.

Temperature changes vary over the globe. Since 1979, land temperatures have increased about twice as fast as ocean temperatures (0.25° C per decade against 0.13° C per decade). Ocean temperatures increase more slowly than land temperatures because of the larger effective heat capacity of the oceans and because the ocean loses more heat by evaporation. The Northern Hemisphere warms faster than the Southern Hemisphere because it has more land and because it has extensive areas of seasonal snow and sea-ice cover subject to ice-albedo feedback. Although more greenhouse gases are emitted in the Northern than Southern Hemisphere this does not contribute to the difference in warming because the major greenhouse gases persist long enough to mix between hemispheres.

The thermal inertia of the oceans and slow responses of other indirect effects mean that climate can take centuries or longer to adjust to changes in forcing. Climate commitment studies indicate that even if greenhouse gases were stabilized at 2000 levels, a further warming of about 0.5° C (0.9° F) would still occur.

*External forcing* refers to processes external to the climate system (though not necessarily external to Earth) that influence climate. Climate responds to several types of external forcing, such as radiative forcing due to changes in atmospheric composition (mainly greenhouse gas concentrations), changes in solar luminosity, volcanic eruptions, and variations in Earth's orbit around the Sun. Attribution of recent climate change focuses on the first three types of forcing. Orbital cycles vary slowly over tens of thousands of years and thus are too gradual to have caused the temperature changes observed in the past century.

## Greenhouse Gases

Recent atmospheric carbon dioxide ($CO_2$) increases. Monthly $CO_2$ measurements display seasonal oscillations in overall yearly uptrend; each year's maximum occurs during the Northern Hemisphere's late spring, and declines during its growing season as plants remove some atmospheric $CO_2$.

The greenhouse effect is the process by which absorption and emission of infrared radiation by gases in the atmosphere warm a planet's lower atmosphere and surface. It was proposed by Joseph Fourier in 1824 and was first investigated quantitatively by Svante Arrhenius in 1896. The question in terms of global warming is how the strength of the presumed greenhouse effect changes when human activity increases the concentrations of greenhouse gases in the atmosphere.

Naturally occurring greenhouse gases have a mean warming effect of about 33° C (59° F). The major greenhouse gases are water vapour, which causes about 36-70 per cent of the greenhouse effect; carbon dioxide ($CO_2$), which causes 9-26 per cent; methane ($CH_4$), which causes 4-9 per cent; and ozone ($O_3$), which causes 3-7 per cent. Clouds also affect the radiation balance, but they are composed of liquid water or ice and so have different effects on radiation from water vapour.

Human activity since the Industrial Revolution has increased the amount of greenhouse gases in the atmosphere, leading to increased radiative forcing from $CO_2$, methane, tropospheric ozone, CFCs and nitrous oxide. The concentrations of $CO_2$ and methane have increased by 36 per cent and 148 per cent respectively since 1750. These levels are much higher than at any time during the last 650,000 years, the period for which reliable data has been extracted from ice cores. Less direct geological evidence indicates that $CO_2$ values higher than this were last seen about 20 million years ago. Fossil fuel burning has produced about three-quarters of the increase in $CO_2$ from human activity over the past 20 years. Most of the rest is due to land-use change, particularly deforestation.

Over the last three decades of the 20th century, GDP per capita and population growth were the main drivers of increases in greenhouse gas emissions. $CO_2$ emissions are continuing to rise due to the burning of fossil fuels and land-use change. 1 Emissions scenarios, estimates of changes in future emission levels of greenhouse gases, have been projected that depend upon uncertain economic, sociological, technological, and natural developments. In most scenarios, emissions continue to rise over the century, while in a few, emissions are reduced. These emission scenarios, combined with carbon cycle modelling, have been used to produce estimates of how atmospheric concentrations of greenhouse gases will change in the future. Using the six IPCC SRES 'marker' scenarios, models suggest that by the year 2100, the atmospheric concentration of $CO_2$ could range between 541 and 970 ppm. This is an increase of 90-250 per cent above the concentration in the year 1750. Fossil fuel reserves are sufficient to reach these levels and continue emissions past 2100 if coal, tar sands or methane clathrates are extensively exploited.

The destruction of stratospheric ozone by chlorofluorocarbons is sometimes mentioned in relation to global warming. Although there are a few areas of linkage, the relationship between the two is not strong. Reduction of stratospheric ozone has a cooling influence. Substantial ozone depletion did not occur until the late 1970s. Ozone in the troposphere (the lowest part of the Earth's atmosphere) does contribute to surface warming.

Global dimming, a gradual reduction in the amount of global direct irradiance at the Earth's surface, has partially counteracted global warming from 1960 to the present. The main cause of this dimming is aerosols produced by volcanoes and pollutants. These aerosols exert a cooling effect by increasing the reflection of incoming sunlight. The effects of the products of fossil fuel combustion–$CO_2$ and aerosols–have largely offset one another in recent decades, so that net warming has been due to the increase in non-$CO_2$ greenhouse gases such as methane. Radiative forcing due to aerosols is temporally limited due to wet deposition which causes aerosols to have an atmospheric lifetime of one week. Carbon dioxide has a lifetime of a century or more, and as such, changes in aerosol concentrations will only delay climate changes due to carbon dioxide.

In addition to their direct effect by scattering and absorbing solar radiation, aerosols have indirect effects on the radiation budget. Sulfate aerosols act as cloud condensation nuclei and thus lead to clouds that have more and smaller cloud droplets. These clouds reflect solar radiation more efficiently than clouds with fewer and larger droplets. This effect also causes droplets to be of more uniform size, which reduces growth of raindrops and makes the cloud more reflective to incoming sunlight. Indirect effects are most noticeable in marine stratiform clouds, and have very little radiative effect on convective clouds. Aerosols, particularly their indirect effects, represent the largest uncertainty in radiative forcing.

Soot may cool or warm the surface, depending on whether it is airborne or deposited. Atmospheric soot aerosols directly absorb solar radiation, which heats the atmosphere and cools the surface. In isolated areas with high soot production, such as rural India, as much as 50 per cent of surface warming due to greenhouse gases may be masked by atmospheric brown clouds. Atmospheric soot always contributes additional warming to the climate system. When deposited, especially on glaciers or on ice in arctic regions, the lower surface albedo can also directly heat the surface. The influences of aerosols, including black carbon, are most pronounced in the tropics and sub-tropics, particularly in Asia, while the effects of greenhouse gases are dominant in the extratropics and southern hemisphere.

Variations in solar output have been the cause of past climate changes. The effect of changes in solar forcing in recent decades is uncertain, but small, with some studies showing a slight cooling effect, while others studies suggest a slight warming effect.

Greenhouse gases and solar forcing affect temperatures in different ways. While both increased solar activity and increased greenhouse gases are expected to warm the troposphere, an increase in solar activity should warm the stratosphere while an increase in greenhouse gases should cool the stratosphere. Observations show that temperatures in the stratosphere have been cooling since 1979, when satellite measurements became available. Radiosonde (weather balloon) data from the pre-satellite era show cooling since 1958, though there is greater uncertainty in the early radiosonde record.

A related hypothesis, proposed by Henrik Svensmark, is that magnetic activity of the sun deflects cosmic rays that may influence the generation of cloud condensation nuclei and thereby affect the climate. Other research has found no relation between warming in recent decades and cosmic rays. The influence of cosmic rays on cloud cover is about a factor of 100 lower than needed to explain the observed changes in clouds or to be a significant contributor to present-day climate change.

Feedback is a process in which changing one quantity changes a second quantity, and the change in the second quantity in turn changes the first.

Positive feedback amplifies the change in the first quantity while negative feedback reduces it. Feedback is important in the study of global warming because it may amplify or diminish the effect of a particular process. The main positive feedback in global warming is the tendency of warming to increase the amount of water vapor in the atmosphere, a significant greenhouse gas. The main negative feedback is radiative cooling, which increases as the fourth power of temperature; the amount of heat radiated from the Earth into space increases with the temperature of Earth's surface and atmosphere. Imperfect understanding of feedbacks is a major cause of uncertainty and concern about global warming.

## Climate Models

Calculations of global warming prepared in or before 2001 from a range of climate models under the SRES A2 emissions scenario, which assumes no action is taken to reduce emissions and regionally divided economic development.

Global warming has been detected in a number of systems. Some of these changes, e.g., based on the instrumental temperature record, have been described in the section on temperature changes. Rising sea levels and observed decreases in snow and ice extent are consistent with warming. Most of the increase in global average temperature since the mid-20th century is, with high probability,[D] atttributable to human-induced changes in greenhouse gas concentrations.

Even with current policies to reduce emissions, global emissions are still expected to continue to grow over the coming decades. Over the course of the 21st century, increases in emissions at or above their current rate would very likely induce changes in the climate system larger than those observed in the 20th century.

In the IPCC Fourth Assessment Report, across a range of future emission scenarios, model-based estimates of sea level rise for the end of the 21st century (the year 2090-2099, relative to 1980-1999) range from 0.18 to 0.59 m. These estimates, however, were not given a likelihood due to a lack of scientific understanding, nor was an upper bound given for sea level rise. Over the course of centuries to millennia, the melting of ice sheets could result in sea level rise of 4-6 m or more.

Changes in regional climate are expected to include greater warming over land, with most warming at high northern latitudes, and least warming over the Southern Ocean and parts of the North Atlantic Ocean. Snow cover area and sea ice extent are expected to decrease. The frequency of hot extremes, heat waves, and heavy precipitation will very likely increase.

## Ecological Systems

In terrestrial ecosystems, the earlier timing of spring events, and poleward and upward shifts in plant and animal ranges, have been linked with high confidence to recent warming. Future climate change is expected to particularly affect certain ecosystems, including tundra, mangroves, and coral reefs. It is expected that most ecosystems will be affected by higher atmospheric $CO_2$ levels, combined with higher global temperatures. Overall, it is expected that climate change will result in the extinction of many species and reduced diversity of ecosystems.

## Social Systems

There is some evidence of regional climate change affecting systems related to human activities, including agricultural and forestry management activities at higher latitudes in the Northern Hemisphere. Future climate change is expected to particularly affect some sectors and systems related to human activities. Low-lying coastal systems are vulnerable to sea level rise and storm surge. Human health will be at increased risk in populations with limited capacity to adapt to climate change. It is expected that some regions will be particularly affected by climate change, including the Arctic, Africa, small islands, and Asian and African megadeltas. In some areas the effects on agriculture, industry and health could be mixed, or even beneficial in certain respects, but overall it is expected that these benefits will be outweighed by negative effects.

The geographic distribution of surface warming during the 21st century calculated by the HadCM3 climate model if a business as usual scenario is assumed for economic growth and greenhouse gas emissions. In this figure, the globally averaged warming corresponds to 3.0° C (5.4° F).

The main tools for projecting future climate changes are mathematical models based on physical principles including fluid dynamics, thermodynamics and radiative transfer. Although they attempt to include as many processes as possible, simplifications of the actual climate system are inevitable because of the constraints of available computer power and limitations in knowledge of the climate system. All modern climate models are in fact *combinations* of models for different parts of the Earth. These include an atmospheric model for air movement, temperature, clouds, and other atmospheric properties; an ocean model that predicts temperature, salt content, and circulation of ocean waters; models for ice cover on land and sea; and a model of heat and moisture transfer from soil and vegetation to the atmosphere. Some models also include treatments of chemical and biological processes. Warming due to increasing levels of greenhouse gases is not an assumption of the models; rather, it is an end result from the interaction of greenhouse gases with radiative transfer and other physical processes. Although much of the variation in model outcomes depends on the greenhouse gas emissions used as inputs, the temperature effect of a specific greenhouse gas concentration (climate sensitivity) varies depending on the model used. The representation of clouds is one of the main sources of uncertainty in present-generation models.

Global climate model projections of future climate most often have used estimates of greenhouse gas emissions from the IPCC Special Report on Emissions Scenarios (SRES). In addition to human-caused emissions, some models also include a simulation of the carbon cycle; this generally shows a positive feedback, though this response is uncertain. Some observational studies also show a positive feedback. Including uncertainties in future greenhouse gas concentrations and climate sensitivity, the IPCC anticipates a warming of 1.1° C to 6.4° C (2.0° F to 11.5° F) by the end of the 21st century, relative to 1980-1999.

Models are also used to help investigate the causes of recent climate change by comparing the observed changes to those that the models project from various natural and human-derived causes. Although these models do not unambiguously attribute the warming that occurred from approximately 1910 to 1945 to either natural variation or human effects, they do indicate that the warming since 1970 is dominated by man-made greenhouse gas emissions.

The physical realism of models is tested by examining their ability to simulate current or past climates. Current climate models produce a good match to observations of global temperature changes over the last century, but do not simulate all aspects of climate. Not all effects of global warming are accurately predicted by the climate models used by the IPCC. Observed Arctic shrinkage has been faster than that predicted. Precipitation increased proportional to atmospheric humidity, and hence significantly faster than current global climate models predict.

Reducing the amount of future climate change is called mitigation of climate change. The IPCC defines mitigation as activities that reduce greenhouse gas (GHG) emissions, or enhance the capacity of carbon sinks to absorb GHGs from the atmosphere. Many countries, both developing and developed, are aiming to use cleaner, less polluting, technologies. 192 Use of these technologies aids mitigation and could result in substantial reductions in $CO_2$ emissions. Policies include targets for emissions reductions, increased use of renewable energy, and increased energy efficiency. Studies indicate substantial potential for future reductions in emissions. Since even in the most optimistic scenario, fossil fuels are going to be used for years to come, mitigation may also involve carbon capture and storage, a process that traps $CO_2$ produced by factories and gas or coal power stations and then stores it, usually underground.

## Adaptation

Other policy responses include adaptation to climate change. Adaptation to climate change may be planned, e.g., by local or national government, or spontaneous, i.e., done privately without government intervention. The ability to adapt is closely linked to social and economic development. Even societies with high capacities to adapt are still vulnerable to climate change. Planned adaptation is already occurring on a limited basis. The barriers, limits, and costs of future adaptation are not fully understood.

Another policy response is engineering of the climate (geoengineering). This policy response is sometimes grouped together with mitigation. Geoengineering is largely unproven, and reliable cost estimates for it have not yet been published.

## UNFCCC

Most countries are Parties to the United Nations Framework Convention on Climate Change (UNFCCC). The ultimate objective of the Convention is to prevent 'dangerous' human interference of the climate system. As is stated in the Convention, this requires that GHGs are stabilized in the atmosphere at a level where ecosystems can adapt naturally to climate change, food production is not threatened, and economic development can proceed in a sustainable fashion.

The UNFCCC recognizes differences among countries in their responsibility to act on climate change. In the Kyoto Protocol to the UNFCCC, most developed countries (listed in Annex I of the treaty) took on legally binding commitments to reduce their emissions. Policy measures taken in response to these commitments have reduced emissions. For many developing (non-Annex I) countries, reducing poverty is their overriding aim.

At the 15th UNFCCC Conference of the Parties, held in 2009 at Copenhagen, several UNFCCC Parties produced the Copenhagen Accord. Parties agreeing with the Accord aim to limit the future increase in global mean temperature to below 2° C.

## Views on Global Warming

There are different views over what the appropriate policy response to climate change should be. These competing views weigh the benefits of limiting emissions of greenhouse gases against the costs. In general, it seems likely that climate change will impose greater damages and risks in poorer regions.

Developing and developed countries have made different arguments over who should bear the burden of costs for cutting emissions. Developing countries often concentrate on per capita emissions, that is, the total emissions of a country divided by its population. Per capita emissions in the industrialized countries are typically as much as ten times the average in developing countries. This is used to make the argument that the real problem of climate change is due to the profligate and unsustainable lifestyles of those living in rich countries. On the other hand, commentators from developed countries more often point out that it is total emissions that matter. In 2008, developing countries made up around half of the world's total emissions of $CO_2$ from cement production and fossil fuel use.

The Kyoto Protocol, which came into force in 2005, sets legally binding emission limitations for most developed countries. Developing countries are not subject to limitations. This exemption led the U.S. and Australia to decide not to ratify the treaty, although Australia did finally ratify the treaty in December 2007.

## Public Opinion

In 2007-2008 Gallup Polls surveyed 127 countries. Over a third of the world's population was unaware of global warming, with people in developing countries less aware than those in developed, and those in Africa the least aware. Of those aware, Latin America leads in belief that temperature changes are a result of human activities while Africa, parts of Asia and the Middle East, and a few countries from the Former Soviet Union lead in the opposite belief. In the Western world, opinions over the concept and the appropriate responses are divided. Nick Pidgeon of Cardiff University said that "results show the different stages of engagement about global warming on each side of the Atlantic", adding, "The debate in Europe is about what action needs to be taken, while many in the U.S. still debate whether climate change is happening".

## Other Views

Most scientists accept that humans are contributing to observed climate change. National science academies have called on world leaders for policies to cut global emissions. However, some scientists and non-scientists question aspects of climate-change science.

Organizations such as the libertarian Competitive Enterprise Institute, conservative commentators, and some companies such as ExxonMobil have challenged IPCC climate change scenarios, funded scientists who disagree with the scientific consensus, and provided their own projections of the economic cost of stricter controls. In the finance industry, Deutsche Bank has set up an institutional climate change investment division DBCCA) which has commissioned and published research on the issues and debate surrounding global warming. Environmental organizations and public figures have emphasized changes in the current climate and the risks they entail, while promoting adaptation to changes in infrastructural needs and emissions reductions. Some fossil fuel companies have scaled back their efforts in recent years, or called for policies to reduce global warming.

The term *global warming* was probably first used in its modern sense on 8 August 1975 in a science paper by Wally Broecker in the journal *Science* called "Are we on the brink of a pronounced global warming?". Broecker's choice of words was new and represented a significant recognition that the climate was warming; previously the phrasing used by scientists was 'inadvertent climate modification', because while it was recognized humans could change the climate, no one was sure which direction it was going. The National Academy of Sciences first used *global warming* in a 1979 paper called the Charney Report, it said: "if carbon dioxide continues to increase, [we find] no reason to doubt that climate changes will result and no reason to believe that these changes will be negligible." The report made a distinction between referring to surface temperature changes as *global warming*, while referring to other changes caused by increased $CO_2$ as *climate change*. This distinction is still often used in science reports, with global warming meaning surface temperatures, and climate change meaning other changes (increased storms, etc.)

*Global warming* became more widely popular after 1988 when NASA scientist James E. Hansen used the term in a testimony to Congress. He said: "global warming has reached a level such that we can ascribe with a high degree of confidence a cause and effect relationship between the greenhouse effect and the observed warming." His testimony was widely reported and afterward *global warming* was commonly used by the press and in public discourse.

CHAPTER 19

# Actinides in the Environment

Actinides in the environment refer to the sources, environmental behaviour and effects of actinides in Earth's environment. Environmental radioactivity is not limited solely to actinides; non-actinides such as radon and radium are of note.

In general for the insoluble actinide oxides such as high fired uranium dioxide and MOX fuel if it is swallowed then it will pass through the digestive system with very little actinide dissolving. As the actinide oxide cannot dissolve, it cannot be absorbed into the body of the person or animal. With such an oxide the dose a person is committed to after a given intake of activity is higher for inhalation than for ingestion as the insoluble compound will remain in the lungs, where it will then irradiate the lung tissue.

Low fired oxides and soluble salts such as the nitrates can be absorbed with greater ease through the digestive system, so they are able to enter the bloodstream after being swallowed. If they are inhaled then it is possible for the solid to dissolve and leave the lungs. Hence the dose to the lungs will be lower for the soluble form.

## Radon and Radium in the Environment

Radon and radium are not actinides—they are both radioactive daughters from the decay of uranium. Aspects of their biology and environmental behaviour is discussed at radium in the environment.

## Thorium in the Environment

In India, a large amount of thorium ore can be found in the form of monazite in placer deposits of the Western and Eastern coastal dune sands, particularly in the Tamil Nadu coastal areas. The residents of this area are exposed to a naturally occurring radiation dose ten times higher than the worldwide average.

Thorium is found in small amounts in most rocks and soils, where it is about three times more abundant than uranium, and is about as common as lead. Soil commonly contains an average of around 6 parts per million (ppm) of thorium. Thorium occurs in several minerals, the most common being the rare earth-thorium-phosphate mineral, monazite, which contains up to about 12 per cent thorium oxide. There are substantial deposits in several countries. $^{232}$Th decays very slowly (its half-life is about three times the age of the earth) but other thorium isotopes occur in the thorium and uranium decay chains. Most of these are short-lived and hence much more radioactive than $^{232}$Th, though on a mass basis they are negligible.

## Effects in Humans

Thorium has been linked to liver cancer. In the past thoria (thorium dioxide) was used as a contrast agent for medical X-ray radiography but its use has been discontinued. It was sold under the name Thorotrast.

## Uranium in the Environment

Uranium is a natural metal which is widely found. It is present in almost all soils and it is more plentiful than antimony, beryllium, cadmium, gold, mercury, silver, or tungsten and is about as abundant as arsenic or molybdenum. Significant concentrations of uranium occur in some substances such as phosphate rock deposits, and minerals such as lignite, and monazite sands in uranium-rich ores (it is recovered commercially from these sources).

Seawater contains about 3.3 parts per billion of uranium by weight as uranium(VI) forms soluble carbonate complexes. The extraction of uranium from seawater has been considered as a means of obtaining the element. Because of the very low specific activity of uranium the chemical effects of it upon living things can often outweigh the effects of its radioactivity. Additional uranium has been added to the environment in some locations as a result of the nuclear fuel cycle and the use of depleted uranium in munitions.

## Neptunium in the Environment

Like plutonium, neptunium has a high affinity for soil. However, it is relatively mobile over the long term, and diffusion of neptunium-237 in groundwater is a major issue in designing a deep geological repository for permanent storage of spent nuclear fuel. $^{237}$Np has a halflife of 2.144 million years, so it is a long-term problem; but its halflife is still much shorter than those of uranium-238, uranium-235, or uranium-236, and $^{237}$Np therefore has higher specific activity than those nuclides.

## Americium in the Environment

Americium often enters landfills from discarded smoke detectors. The rules associated with the disposal of smoke detectors are very relaxed in most municipalities. For instance in the UK it is permissible to dispose of an americium containing smoke detector by placing it in the dustbin with normal household rubbish, but each dustbin worth of rubbish is limited to only containing one smoke detector.

In France a truck transporting 900 smoke detectors had been reported to have caught fire, it is claimed that this led to a release of americium into the environment. In the U.S., the 'Radioactive Boy Scout' David Hahn was able to buy thousands of smoke detectors at remainder prices and concentrate the americium from them.

There have been cases of humans being contaminated with americium, the worst case being that of Harold McCluskey. It is interesting to note that Harold McCluskey did not die of cancer but of heart disease (which he had before the accident). It is likely that the medical care which he was given saved his life; because of the difference in the chemistry of americium (the +3 oxidation state is very stable) to plutonium (where the +4 state can form in the human body) the americium has very different biochemistry to plutonium.

The most common isotope americium-241 decays (halflife 431 years) to neptunium-237 which has a much longer halflife, so in the long term, the issues discussed above for neptunium apply.

## Sources

Plutonium in the environment has several sources. These include:

- Atomic batteries
- In space
- In pacemakers
- Bomb detonations
- Bomb safety trials
- Nuclear accidents (such as Chernobyl)
- Nuclear crime
- Nuclear fuel cycle

## Environmental Chemistry

Plutonium, like other actinides, readily forms a plutonium dioxide (*plutonyl*) core ($PuO_2$). In the environment, this plutonyl core readily complexes with carbonate as well as other oxygen moieties ($OH^-$, $NO_2^-$, $NO_3$, and $SO_4^{-2}$) to form charged complexes which can be readily mobile with low affinities to soil.

- $PuO_2(CO_3)_1^{-2}$
- $PuO_2(CO_3)_2^{-4}$
- $PuO_2(CO_3)_3^{-6}$

$PuO_2$ formed from neutralizing highly acidic nitric acid solutions tends to form polymeric $PuO_2$ which is resistant to complexation. Plutonium also readily shifts valences between the +3, +4, +5 and +6 states. It is common for some fraction of plutonium in solution to exist in all of these states in equilibrium.

Plutonium is known to bind to soil particles very strongly, see above for a X-ray spectrscopic study of plutonium in soil and concrete. While caesium has very different chemistry to the actinides, it is well known that both caesium and many of the actinides bind strongly to the minerals in soil. Hence it has been possible to use $^{134}Cs$ labeled soil to study the migration of Pu and Cs is soils. It has been shown that colloidal transport processes control the migration of Cs (and will control the migration of Pu) in the soil at the Waste Isolation Pilot Plant.

## Environmental Conservation

Environmental conservation is the process in which one is involved in conserving the natural aspects of the environment. Whether through reforestation, recycling, or pollution control, environmental conservation sustains the natural quality of life. Environmental health movement dates at least to Progressive Era, and focuses on urban standards like clean water, efficient sewage handling, and stable population growth. Environmental health could also deal with nutrition, preventive medicine, aging, and other concerns specific to human well-being. Environmental health is also seen as an indicator for the state of the environment, or an early warning system for what may happen to humans Environmental Justice is a movement that began in the U.S. in the 1980s and seeks an end to environmental racism and prevent low-income and minority communities from an unbalanced exposure to highways, garbage dumps, and factories. The Environmental Justice movement seeks to link 'social' and 'ecological' environmental concerns, while at the same time preventing de facto racism, and classism. This makes it particularly adequate for the construction of labour-environmental alliances. Ecology movement could involve the Gaia Theory, as well as Value of Earth and other interactions between humans, science, and responsibility. Deep Ecology is an ideological spinoff of the ecology movement that views the diversity and integrity of the planetary ecosystem, in and for itself, as its primary value. Bright green environmentalism is a currently popular sub-movement, which emphasizes the idea that through technology, good design and more thoughtful use of energy and resources, people can live responsible, sustainable lives while enjoying prosperity. The anti-nuclear movement opposes the use of various nuclear technologies. The initial anti-nuclear objective was nuclear disarmament and later the focus began to shift to other issues, mainly opposition to the use of nuclear power. There have been many large anti-nuclear demonstrations and protests. Major anti-nuclear groups include Campaign for Nuclear Disarmament, Friends of the Earth, Greenpeace, International Physicians for the Prevention of Nuclear War, and the Nuclear Information and Resource Service.

Today, the sciences of ecology and environmental science, rather than any aesthetic goals, provide the basis of unity to most serious environmentalists. As more information is gathered in scientific fields, more scientific issues like biodiversity, as opposed to mere aesthetics, are a concern. Conservation biology is a rapidly developing field. Environmentalism now has proponents in business: new ventures such as those to reuse and recycle consumer electronics and other technical equipment are gaining popularity. Computer liquidators are just one example.

In recent years, the environmental movement has increasingly focused on global warming as a top issue. As concerns about climate change moved more into the mainstream, from the connections drawn between global warming and Hurricane Katrina to Al Gore's film An Inconvenient Truth, many environmental groups refocussed their efforts. In the United States, 2007 witnessed the largest grassroots environmental demonstration in years, Step It Up 2007, with rallies in over 1,400 communities and all 50 states for real global warming solutions.

Many religious organizations and individual churches now have programmes and activities dedicated to environmental issues. The religious movement is often supported by interpretation of scriptures. Most major religious groups are represented including Jewish, Islamic, Anglican, Orthodox, Evangelical, Christian and Catholic.

Radical environmentalism emerged out of an ecocentrism-based frustration with the co-option of mainstream environmentalism. The radical environmental movement aspires to what scholar Christopher Manes calls "a new kind of environmental activism: iconoclastic, uncompromising, discontented with traditional conservation policy, at time illegal ..." Radical environmentalism presupposes a need to reconsider Western ideas of religion and philosophy (including capitalism, patriarchy and globalization) sometimes through 'resacralising' and reconnecting with nature. Greenpeace represents an organisation with a radical approach, but has contributed in serious ways towards understanding of critical issues, and has a science-oriented core with radicalism as a means to mediaexposure. Earth-first, takes a much more radical posture. Greenpeace, Earth-first.

A concern for environmental protection has recurred in diverse forms, in different parts of the world, throughout history. For example, in the Middle East, the earliest known writings concerned with environmental pollution were Arabic medical treatises written during the 'Arab Agricultural Revolution', by writers such as Alkindus, Costa ben Luca, Rhazes, Ibn Al-Jazzar, al-Tamimi, al-Masihi, Avicenna, Ali ibn Ridwan, Isaac Israeli ben Solomon, Abd-el-latif, and Ibn al-Nafis. They were concerned with air contamination, water contamination, soil contamination, solid waste mishandling, and environmental assessments of certain localities.

In Europe, King Edward I of England banned the burning of sea-coal by proclamation in London in 1272, after its smoke had become a problem. The fuel was so common in England that this earliest of names for it was acquired because it could be carted away from some shores by the wheelbarrow. Air pollution would continue to be a problem in England, especially later during the Industrial Revolution, and extending into the recent past with the Great Smog of 1952.

## Origins of the Modern Environmental Movement

In Europe, the Industrial Revolution gave rise to modern environmental pollution as it is generally understood today. The emergence of great factories and consumption of immense quantities of coal and other fossil fuels gave rise to unprecedented air pollution and the large volume of industrial chemical discharges added to the growing load of untreated human waste. The first large-scale, modern environmental laws came in the form of the British Alkali Acts, passed in 1863, to regulate the deleterious air pollution (gaseous hydrochloric acid) given off by the Leblanc process, used to produce soda ash. Environmentalism grew out of the amenity movement, which was a reaction to industrialization, the growth of cities, and worsening air and water pollution.

In Victorian Britain, an early 'Back-to-Nature' movement that anticipated modern environmentalism was advocated by intellectuals such as John Ruskin, William Morris and Edward Carpenter, who were all against consumerism, pollution and other activities that were harmful to the natural world. Their ideas also inspired various proto-environmental groups in the UK, such as the Commons Preservation Society, the Kyrle

Society, the Royal Society for the Protection of Birds and the Garden city movement, as well as encouraging the Socialist League and The Clarion movement to advocate measures of nature conservation.

In the United States, the beginnings of an environmental movement can be traced as far back as 1739, though it was not called environmentalism and was still considered conservation until the 1950s. Benjamin Franklin and other Philadelphia residents, citing 'public rights', petitioned the Pennsylvania Assembly to stop waste dumping and remove tanneries from Philadelphia's commercial district. The US movement expanded in the 1800s, out of concerns for protecting the natural resources of the West, with individuals such as John Muir and Henry David Thoreau making key philosophical contributions. Thoreau was interested in peoples' relationship with nature and studied this by living close to nature in a simple life. He published his experiences in the book *Walden,* which argues that people should become intimately close with nature. Muir came to believe in nature's inherent right, especially after spending time hiking in Yosemite Valley and studying both the ecology and geology. He successfully lobbied congress to form Yosemite National Park and went on to set up the Sierra Club. The conservationist principles as well as the belief in an inherent right of nature were to become the bedrock of modern environmentalism.

In the 20th century, environmental ideas continued to grow in popularity and recognition. Efforts were starting to be made to save some wildlife, particularly the American Bison. The death of the last Passenger Pigeon as well as the endangerment of the American Bison helped to focus the minds of conservationists and popularize their concerns. In 1916 the National Park Service was founded by US President Woodrow Wilson.

In 1949, *A Sand County Almanac* by Aldo Leopold was published. It explained Leopold's belief that humankind should have moral respect for the environment and that it is unethical to harm it. The book is sometimes called the most influential book on conservation.

Throughout the 1950s, 1960s, 1970s and beyond, photography was used to enhance public awareness of the need for protecting land and recruiting members to environmental organizations. David Brower, Ansel Adams and Nancy Newhall created the Sierra Club Exhibit Format Series, which helped raise public environmental awareness and brought a rapidly increasing flood of new members to the Sierra Club and to the environmental movement in general. 'This Is Dinosaur' edited by Wallace Stegner with photographs by Martin Litton and Philip Hyde prevented the building of dams within Dinosaur National Monument by becoming part of a new kind of activism called environmentalism that combined the conservationist ideals of Thoreau, Leopold and Muir with hard-hitting advertising, lobbying, book distribution, letter writing campaigns, and more. The powerful use of photography in addition to the written word for conservation dated back to the creation of Yosemite National Park, when photographs convinced Abraham Lincoln to preserve the beautiful glacier carved landscape for all time. The Sierra Club Exhibit Format Series galvanized public opposition to building dams in the Grand Canyon and protected many other national treasures. The Sierra Club often led a coalition of many environmental groups including the Wilderness Society and many others. After a focus on preserving wilderness in the 1950s and 1960s, the Sierra Club and other groups broadened their focus to include such issues as air and water pollution, population control, and curbing the exploitation of natural resources.

In 1962, *Silent Spring* by American biologist Rachel Carson was published. The book cataloged the environmental impacts of the indiscriminate spraying of DDT in the US and questioned the logic of releasing large amounts of chemicals into the environment without fully understanding their effects on ecology or human health. The book suggested that DDT and other pesticides may cause cancer and that their agricultural use was a threat to wildlife, particularly birds. The resulting public concern led to the creation of the United States Environmental Protection Agency in 1970 which subsequently banned the agricultural use of DDT in the US in 1972. The limited use of DDT in disease vector control continues to this day in certain parts of the world and remains controversial. The book's legacy was to produce a far greater awareness of environmental issues and interest into how people affect the environment. With this new interest in environment came interest in problems such as air pollution and petroleum spills, and environmental interest grew. New pressure groups formed, notably Greenpeace and Friends of the Earth.

In the 1970s, the Chipko movement was formed in India; influenced by Mohandas Gandhi, they set up peaceful resistance to deforestation by literally hugging trees (leading to the term 'tree huggers'). Their peaceful methods of protest and slogan 'ecology is permanent economy' were very influential.

By the mid-1970s, many felt that people were on the edge of environmental catastrophe. The Back-to-the-land movement started to form and ideas of environmental ethics joined with anti-Vietnam War sentiments and other political issues. These individuals lived outside normal society and started to take on some of the more radical environmental theories such as deep ecology. Around this time more mainstream environmentalism was starting to show force with the signing of the Endangered Species Act in 1973 and the formation of CITES in 1975.

In 1979, James Lovelock, a former NASA scientist, published *Gaia: A new look at life on Earth*, which put forth the Gaia Hypothesis; it proposes that life on Earth can be understood as a single organism. This became an important part of the Deep Green ideology. Throughout the rest of the history of environmentalism there has been debate and argument between more radical followers of this Deep Green ideology and more mainstream environmentalists.

## Environmental Organizations and Conferences

Environmental organizations can be global, regional, national or local; they can be government-run or private (NGO). Environmentalist activity exists in almost every country. Moreover, groups dedicated to community development and social justice also focus on environmental concerns.

There are some volunteer organizations. For example Ecoworld, which is about the environment and is based in team work and volunteer work. Some US environmental organizations, among them the Natural Resources Defense Council and the Environmental Defense Fund, specialize in bringing lawsuits (a tactic seen as particularly useful in that country). Other groups, such as the US-based National Wildlife Federation, the Nature Conservancy, and The Wilderness Society, and global groups like the World Wide Fund for Nature and Friends of the Earth, disseminate information, participate in public hearings, lobby, stage demonstrations, and may purchase land for preservation. Smaller groups, including Wildlife Conservation International, conduct research on endangered species and ecosystems. More radical organizations, such as Greenpeace, Earth First!, and the Earth Liberation Front, have more directly opposed actions they regard as environmentally harmful. While Greenpeace is devoted to nonviolent confrontation as a means of bearing witness to environmental wrongs and bringing issues into the public realm for debate, the underground Earth Liberation Front engages in the clandestine destruction of property, the release of caged or penned animals, and other criminal acts. Such tactics are regarded as unusual within the movement, however.

On an international level, concern for the environment was the subject of a UN conference in Stockholm in 1972, attended by 114 nations. Out of this meeting developed UNEP (United Nations Environment Programme) and the follow-up United Nations Conference on Environment and Development in 1992. Other international organizations in support of environmental policies development include the Commission for Environmental Cooperation (NAFTA), the European Environment Agency (EEA), and the Intergovernmental Panel on Climate Change.

# Index